STUDENT'S SOLUTIONS MANUAL

EDGAR N. REYES

Southeastern Louisiana University

FUNDAMENTALS OF PRECALCULUS

SECOND EDITION

Mark Dugopolski

Southeastern Louisiana University

PEARSON

Addison
Wesley

Boston San Francisco New York
London Toronto Sydney Tokyo Singapore Madrid
Mexico City Munich Paris Cape Town Hong Kong Montreal

Reproduced by Pearson Addison-Wesley from electronic files supplied by the author.

Copyright © 2009 Pearson Education, Inc.
Publishing as Pearson Addison-Wesley, 75 Arlington Street, Boston, MA 02116.

ISBN-13: 978-0-321-53662-4
ISBN-10: 0-321-53662-2

7 V036 10

This Student's Solutions Manual includes worked out solutions to all odd-numbered exercises in the exercise sets, and all odd-numbered exercises in the end of the chapter "Review Exercises". Also, included are solutions to the "For Thought", "Chapter Test", and "Tying It All Together" exercises.

TABLE OF CONTENTS

Chapter 1 Graphs and Functions   1

Chapter 2 Polynomial and Rational Functions   36

Chapter 3 Trigonometric Functions   69

Chapter 4 Exponential and Logarithmic Functions   106

Chapter 5 Conic Sections, Polar Coordinates, and Parametric Equations 122

Appendix Basic Algebra Review   149

For Thought

1. False, 0 is not an irrational number.

2. False, since 0 has no multiplicative inverse.

3. True

4. False, since the statement fails when $a = w = z$.

5. False, since $a - (b - c) = a - b + c$.

6. False, the distance is $|a - b|$.

7. False, a TI-92 can give an exact answer like π.

8. False, the opposite of $a + b$ is $-a - b$.

9. True, since $|x - 9| = 2$ is equivalent to

$$\begin{array}{rclcrcl} x - 9 & = & 2 & \text{or} & x - 9 & = & -2 \\ x & = & 2 + 9 & \text{or} & x & = & -2 + 9 \\ x & = & 11 & \text{or} & x & = & 7 \end{array}$$

Then the solution set is $\{7, 11\}$.

10. True, since $|x - 5|$ is a nonnegative number for any x.

1.1 Exercises

1. e, true since $\sqrt{2}$ is an irrational number and all irrational numbers are real numbers.

3. h, true since 0 is a rational number.

5. g, true since each integer is a real number.

7. c, true since an irrational number is not an element of the set of rational numbers.

9. All

11. $\{-\sqrt{2}, \sqrt{3}, \pi, 5.090090009...\}$

13. $\{0, 1\}$ **15.** $x + 7$ **17.** $5x + 15$

19. $5(x + 1)$ **21.** $(-13 + 4) + x$

23. $\dfrac{1}{0.125} = 8$ **25.** $\sqrt{3}$

27. $y^2 - x^2$ **29.** 7.2 **31.** $\sqrt{5}$

33. $|13 - 8| = 5$

35. $|17 - (-5)| = 22$

37. $|-18 - (-6)| = |-12| = 12$

39. $\left| \dfrac{1}{4} - \left(-\dfrac{1}{2} \right) \right| = \left| \dfrac{3}{4} \right| = \dfrac{3}{4}$

41. Note, an equation is $|x - 7| = 3$.

$$\begin{array}{rclcrcl} x - 7 & = & 3 & \text{or} & x - 7 & = & -3 \\ x & = & 3 + 7 & \text{or} & x & = & -3 + 7 \\ x & = & 10 & \text{or} & x & = & 4 \end{array}$$

Then the solution set is $\{4, 10\}$.

43. Note, an equation is $|-1 - x| = 3$.

$$\begin{array}{rclcrcl} -1 - x & = & 3 & \text{or} & -1 - x & = & -3 \\ -1 - 3 & = & x & \text{or} & -1 + 3 & = & x \\ -4 & = & x & \text{or} & 2 & = & x \end{array}$$

Then the solution set is $\{-4, 2\}$.

45. Note, an equation is $|x - (-9)| = 4$.

$$\begin{array}{rclcrcl} x + 9 & = & 4 & \text{or} & x + 9 & = & -4 \\ x & = & 4 - 9 & \text{or} & x & = & -4 - 9 \\ x & = & -5 & \text{or} & x & = & -13 \end{array}$$

Then the solution set is $\{-13, -5\}$.

47. $\{\pm 9\}$

49. Since $5x - 4 = 0$, we find $x = \dfrac{4}{5}$.

The solution set is $\left\{ \dfrac{4}{5} \right\}$.

51. Since $2x - 3 = 7$ or $2x - 3 = -7$, we get $2x = 10$ or $2x = -4$.
The solution set is $\{5, -2\}$.

53. Dividing $2|x + 5| = 10$ by 2 we obtain $|x + 5| = 5$.
Then $x + 5 = 5$ or $x + 5 = -5$.
The solution set is $\{0, -10\}$.

55. By dividing by 8, we get $|3x - 2| = \dfrac{0}{8} = 0$.

Since $3x - 2 = 0$, the solution set is $\left\{ \dfrac{2}{3} \right\}$.

57. Solving for $|x|$, we find $|x| = -\dfrac{1}{2}$.
The equation has no solution

59. Arranged from smallest to largest, we get

$$-\frac{1}{2}, -\frac{5}{12}, -\frac{1}{3}, 0, \frac{1}{3}, \frac{5}{12}, \frac{1}{2}.$$

To compare the numbers , rewrite each number as a fraction with denominator 12.

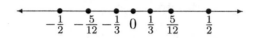

For Thought

1. True

2. False, since $-2x < -6$ is equivalent to

$$\frac{-2x}{-2} > \frac{-6}{-2}.$$

3. False, since there is a number between any two distinct real numbers.

4. True, since $|-6-6| = |-12| = 12 > -1.$

5. False, $(-\infty, -3) \cap (-\infty, -2) = (-\infty, -3).$

6. False, $(5, \infty) \cap (-\infty, -3) = \phi.$

7. False, no real number satisfies $|x - 2| < 0.$

8. False, it is equivalent to $|x| > 3.$

9. False, $|x| + 2 < 5$ is equivalent to $-3 < x < 3.$

10. True

1.2 Exercises

1. $x < 12$

3. $x \geq -7$

5. $[-8, \infty)$

7. $(-\infty, \pi/2)$

9. Since $3x > 15$ implies $x > 5$, the solution set is $(5, \infty)$ and the graph is

11. Since $10 \leq 5x$ implies $2 \leq x$, the solution set is $[2, \infty)$ and the graph is

13. Multiply 6 to both sides of the inequality.

$$
\begin{aligned}
3x - 24 \;&<\; 2x + 30 \\
x \;&<\; 54
\end{aligned}
$$

The solution is the interval $(-\infty, 54)$ and the graph is

15. Multiplying the inequality by 2, we find

$$
\begin{aligned}
7 - 3x \;&\geq\; -6 \\
13 \;&\geq\; 3x \\
13/3 \;&\geq\; x.
\end{aligned}
$$

The solution is the interval $(-\infty, 13/3]$ and the graph is

17. Multiply the inequality by -5 and reverse the direction of the inequality.

$$
\begin{aligned}
2x - 3 \;&\leq\; 0 \\
2x \;&\leq\; 3 \\
x \;&\leq\; \frac{3}{2}
\end{aligned}
$$

The solution is the interval $(-\infty, 3/2]$ and the graph is

19. Multiply the left-hand side.

$$
\begin{aligned}
-6x + 4 \;&\geq\; 4 - x \\
0 \;&\geq\; 5x \\
0 \;&\geq\; x.
\end{aligned}
$$

The solution is the interval $(-\infty, 0]$ and the graph is

21. $(-3, \infty)$

23. $(-3, \infty)$

25. $(-5, -2)$

27. ϕ

29. $(-\infty, 5]$

31. Solve each simple inequality and find the intersection of their solution sets.

$$x > 3 \quad \text{and} \quad 0.5x < 3$$
$$x > 3 \quad \text{and} \quad x < 6$$

The intersection of these values of x is the interval $(3, 6)$ and whose graph is

3   6

33. Solve each simple inequality and find the intersection of their solution sets.

$$2x - 5 > -4 \quad \text{and} \quad 2x + 1 > 0$$
$$x > \frac{1}{2} \quad \text{and} \quad x > -\frac{1}{2}$$

The intersection of these values of x is the interval $(1/2, \infty)$ and the graph is

1/2

35. Solve each simple inequality and find the union of their solution sets.

$$-6 < 2x \quad \text{or} \quad 3x > -3$$
$$-3 < x \quad \text{or} \quad x > -1$$

The union of these values of x is $(-3, \infty)$ and the graph is

−3

37. Solve each simple inequality and find the union of their solution sets.

$$x + 1 > 6 \quad \text{or} \quad x < 7$$
$$x > 5 \quad \text{or} \quad x < 7$$

The union of these values of x is $(-\infty, \infty)$ and the graph is

39. Solve each simple inequality and find the intersection of their solution sets.

$$2 - 3x < 8 \quad \text{and} \quad x - 8 \le -12$$
$$-6 < 3x \quad \text{and} \quad x \le -4$$
$$-2 < x \quad \text{and} \quad x \le -4$$

The intersection is empty and there is no solution.

41. Solve each simple inequality. Then find the intersection of the solution sets.

$$1 < 3x - 5 \quad \text{and} \quad 3x - 5 < 7$$
$$6 < 3x \quad \text{and} \quad 3x < 12$$
$$2 < x \quad \text{and} \quad x < 4$$

The intersection of these values of x is the interval $(2, 4)$ and the graph is

2   4

43. Solve each simple inequality. Then find the intersection of the solution sets.

$$-2 \le 4 - 6x \quad \text{and} \quad 4 - 6x < 22$$
$$6x \le 6 \quad \text{and} \quad -18 < 6x$$
$$x \le 1 \quad \text{and} \quad -3 < x$$

The intersection of these values of x is the interval $(-3, 1]$ and the graph is

-3   1

45. Solve an equivalent compound inequality.

$$-2 < 3x - 1 < 2$$
$$-1 < 3x < 3$$
$$-\frac{1}{3} < x < 1$$

The solution set is the interval $(-1/3, 1)$ and the graph is

-1/3   1

47. Solve an equivalent compound inequality.

$$-1 \le 5 - 4x \le 1$$
$$-6 \le -4x \le -4$$
$$\frac{3}{2} \ge x \ge 1$$

The solution set is the interval $[1, 3/2]$ and the graph is

1   3/2

49. Solve an equivalent compound inequality.

$$-5 \le 4 - x \le 5$$
$$-9 \le -x \le 1$$
$$9 \ge x \ge -1$$

The solution set is the interval $[-1, 9]$ and the graph is

-1   9

51. No solution since an absolute value is never negative.

53. No solution since an absolute value is never negative.

55. Note, $3|x - 2| > 3$ or $|x - 2| > 1$.
We solve an equivalent compound inequality.

$$x - 2 > 1 \quad \text{or} \quad x - 2 < -1$$
$$x > 3 \quad \text{or} \quad x < 1$$

The solution set is the interval $(-\infty, 1) \cup (3, \infty)$ and the graph is

57. Solve an equivalent compound inequality.

$$\frac{x - 3}{2} > 1 \quad \text{or} \quad \frac{x - 3}{2} < -1$$
$$x - 3 > 2 \quad \text{or} \quad x - 3 < -2$$
$$x > 5 \quad \text{or} \quad x < 1$$

The solution set is the interval $(-\infty, 1) \cup (5, \infty)$ and the graph is

59. $|x| < 5$

61. $|x| > 3$

63. Since 6 is the midpoint of 4 and 8, the inequality is
$$|x - 6| < 2.$$

65. Since 4 is the midpoint of 3 and 5, the inequality is
$$|x - 4| > 1.$$

67. $|x| \geq 9$

69. Since 7 is the midpoint, the inequality is $|x - 7| \leq 4$.

71. Since 5 is the midpoint, the inequality is $|x - 5| > 2$.

73. Let x be Lucky's score on the final exam.

$$79 \quad < \quad \frac{65 + x}{2} < 90$$
$$158 \quad < \quad 65 + x < 180$$
$$93 \quad < \quad x < 115.$$

The final exam score must lie in $(93, 115)$.

75. Let x be Ingrid's final exam score. Since $\dfrac{2x + 65}{3}$ is her weighted average, we obtain

$$79 \quad < \quad \frac{2x + 65}{3} < 90$$
$$237 \quad < \quad 2x + 65 < 270$$
$$172 \quad < \quad 2x < 205$$
$$86 \quad < \quad x < 102.5$$

Ingrid's final exam score must lie in $(86, 102.5)$.

77. If x is the price of a car excluding sales tax then it must satisfy $0 \leq 1.1x + 300 \leq 8000$. This is equivalent to $0 \leq x \leq \dfrac{7700}{1.1} = 7000$. The price range of Yolanda's car is the interval $[\$0, \$7000]$.

79. By substituting $N = 50$ and $w = 27$ into

$$r = \frac{Nw}{n}$$

we find

$$r = \frac{1350}{n}.$$

Moreover if $n = 14$, then $r = \dfrac{1350}{14} = 96.4 \approx 96$. Similarly, the other gear ratios are the following:

n	14	17	20	24	29
r	96	79	68	56	47

Yes, the bicycle has a gear ratio for each of the four types.

81. If x is the price of a BMW 760 Li, then $|x - 74,595| > 25,000$. An equivalent inequality is

$$x - 74,595 > 25,000 \quad \text{or} \quad x - 74,595 < -25,000$$
$$x > 99,595 \quad \text{or} \quad x < 49,595.$$

The price of a BMW 760 Li is either under $\$49,595$ or over $\$99,595$.

83. If x is the actual temperature, then

$$\left| \frac{x - 35}{35} \right| \quad < \quad .01$$

$$-.35 < x - 35 \quad < \quad .35$$
$$34.65 < x \quad < \quad 35.35.$$

The actual temperature must lie in the interval $(34.65°, 35.35°)$.

85. If c is the actual circumference, then $c = \pi d$ and

$$
\begin{aligned}
|\pi d - 7.2| &\leq 0.1 \\
-0.1 \leq \pi d - 7.2 &\leq 0.1 \\
7.1 \leq \pi d &\leq 7.3 \\
2.26 \leq d &\leq 2.32.
\end{aligned}
$$

The actual diameter must lie in the interval $[2.26 \text{ cm}, 2.32 \text{ cm}]$.

For Thought

1. False, the point $(2, -3)$ is in Quadrant IV.

2. False, the point $(4, 0)$ does not belong to any quadrant.

3. False, since the distance is $\sqrt{(a-c)^2 + (b-d)^2}$.

4. False, since $Ax + By = C$ is a linear equation.

5. True, since the x-intercept can be obtained by replacing y by 0.

6. False, since $\sqrt{7^2 + 9^2} = \sqrt{130} \approx 11.4$

7. True

8. True

9. True

10. False, it is a circle of radius $\sqrt{5}$.

1.3 Exercises

1. $(4, 1)$, Quadrant I

3. $(1, 0)$, x-axis

5. $(5, -1)$, Quadrant IV

7. $(-4, -2)$, Quadrant III

9. $(-2, 4)$, Quadrant II

11. Distance is $\sqrt{(4-1)^2 + (7-3)^2} = \sqrt{9 + 16} = \sqrt{25} = 5$, midpoint is $(2.5, 5)$

13. Distance is $\sqrt{(-1-1)^2 + (-2-0)^2} = \sqrt{4+4} = 2\sqrt{2}$, midpoint is $(0, -1)$

15. Distance is $\sqrt{(-1 + 3\sqrt{3} - (-1))^2 + (4-1)^2} = \sqrt{27 + 9} = 6$, midpoint is $\left(\dfrac{-2 + 3\sqrt{3}}{2}, \dfrac{5}{2}\right)$

17. Distance is $\sqrt{(1.2 + 3.8)^2 + (4.4 + 2.2)^2} = \sqrt{25 + 49} = \sqrt{74}$, midpoint is $(-1.3, 1.3)$

19. Distance is $\sqrt{(a-b)^2 + 0} = |a - b|$, midpoint is $\left(\dfrac{a+b}{2}, 0\right)$

21. Distance is $\dfrac{\sqrt{\pi^2 + 4}}{2}$, midpoint is $\left(\dfrac{3\pi}{4}, \dfrac{1}{2}\right)$

23. Center $(0, 0)$, radius 4

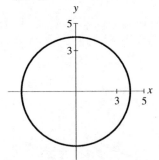

25. Center $(-6, 0)$, radius 6

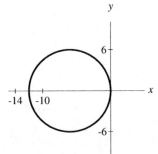

27. Center $(-1, 0)$, radius 5

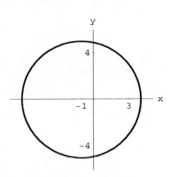

29. Center $(2, -2)$, radius $2\sqrt{2}$

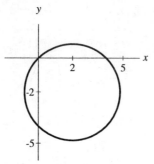

31. $x^2 + y^2 = 49$

33. $(x + 2)^2 + (y - 5)^2 = 1/4$

35. The distance between $(3, 5)$ and the origin is $\sqrt{34}$ which is the radius. The standard equation is $(x - 3)^2 + (y - 5)^2 = 34$.

37. The distance between $(5, -1)$ and $(1, 3)$ is $\sqrt{32}$ which is the radius. The standard equation is $(x - 5)^2 + (y + 1)^2 = 32$.

39. Center $(0, 0)$, radius 3

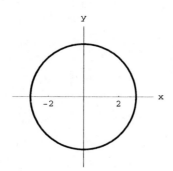

41. Completing the square, we have

$$
\begin{aligned}
x^2 + (y^2 + 6y + 9) &= 0 + 9 \\
x^2 + (y + 3)^2 &= 9.
\end{aligned}
$$

The center is $(0, -3)$ and the radius is 3.

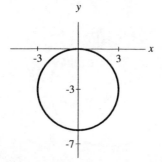

43. Completing the square, we obtain

$$
\begin{aligned}
(x^2 - 6x + 9) + (y^2 - 8y + 16) &= 9 + 16 \\
(x - 3)^2 + (y - 4)^2 &= 25.
\end{aligned}
$$

The center is $(3, 4)$ and the radius is 5.

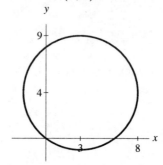

45. Completing the square, we obtain

$$
\begin{aligned}
(x^2 - 4x + 4) + \left(y^2 - 3y + \frac{9}{4}\right) &= 4 + \frac{9}{4} \\
(x - 2)^2 + \left(y - \frac{3}{2}\right)^2 &= \frac{25}{4}.
\end{aligned}
$$

The center is $\left(2, \frac{3}{2}\right)$ and the radius is $\frac{5}{2}$.

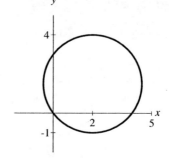

47. Completing the square, we obtain

$$
\begin{aligned}
\left(x^2 - \frac{1}{2}x + \frac{1}{16}\right) + \left(y^2 + \frac{1}{3}y + \frac{1}{36}\right) &= \frac{1}{36} \\
\left(x - \frac{1}{4}\right)^2 + \left(y + \frac{1}{6}\right)^2 &= \frac{1}{36}.
\end{aligned}
$$

The center is $\left(\frac{1}{4}, -\frac{1}{6}\right)$ and the radius is $\frac{1}{6}$.

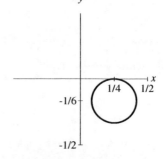

49. a) Since the center is $(0,0)$ and the radius is 7, the standard equation is $x^2 + y^2 = 49$.

b) The radius, which is the distance between $(1,0)$ and $(3,4)$, is given by

$$\sqrt{(3-1)^2 + (4-0)^2} = \sqrt{20}.$$

Together with the center $(1,0)$, it follows that the standard equation is

$$(x-1)^2 + y^2 = 20.$$

c) The center is

$$\left(\frac{3-1}{2}, \frac{5-1}{2}\right) = (1,2)$$

by the midpoint formula. The diameter is

$$\sqrt{(3-(-1))^2 + (5-(-1))^2} = \sqrt{52}.$$

Since the square of the radius is $\left(\sqrt{52}/2\right)^2 = 13$, the equation is

$$(x-1)^2 + (y-2)^2 = 13.$$

51. a. Since the center is $(2,-3)$ and the radius is 2, the standard equation is

$$(x-2)^2 + (y+3)^2 = 4.$$

b. The center is $(-2,1)$, the radius is 1, and the standard equation is

$$(x+2)^2 + (y-1)^2 = 1.$$

c. The center is $(3,-1)$, the radius is 3, and the standard equation is

$$(x-3)^2 + (y+1)^2 = 9.$$

d. The center is $(0,0)$, the radius is 1, and the standard equation is

$$x^2 + y^2 = 1.$$

53. $y = 3x - 4$ goes through $(0,-4)$, $\left(\frac{4}{3}, 0\right)$.

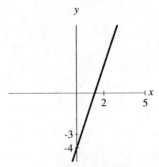

55. $3x - y = 6$ goes through $(0,-6)$, $(2,0)$.

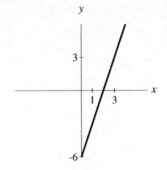

57. $x = 3y - 90$ goes through $(0,30)$, $(-90,0)$.

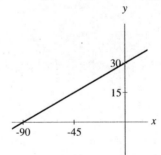

59. $\frac{2}{3}y - \frac{1}{2}x = 400$ goes through $(0,600)$, $(-800,0)$.

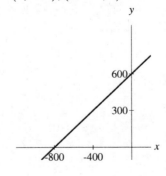

61. Intercepts are $(0, 0.0025), (0.005, 0)$.

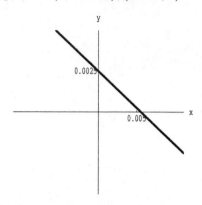

63. Intercepts are $(0, 2500), (5000, 0)$.

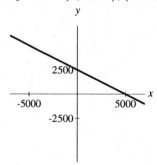

65. $x = 5$

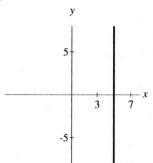

67. $y = 4$

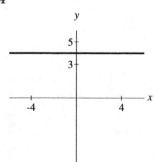

69. $x = -4$

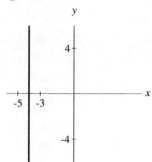

71. Solving for y, we have $y = 1$.

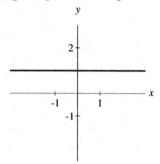

73. Since the x-intercept of $y = 2.4x - 8.64$ is $(3.6, 0)$, the solution set of $2.4x - 8.64 = 0$ is $\{3.6\}$.

75. Since the x-intercept of $y = -\dfrac{3}{7}x + 6$ is $(14, 0)$, the solution set of $-\dfrac{3}{7}x + 6 = 0$ is $\{14\}$.

77. The solution is $x = -\dfrac{3.4}{12} \approx -2.83$.

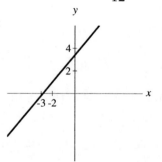

79. The solution is $\dfrac{3497}{0.03} \approx 116,566.67$

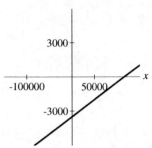

81. Solve for x.

$$4.3 - 3.1(2.3x) + 3.1(9.9) = 0$$
$$4.3 - 7.13x + 30.69 = 0$$
$$34.99 - 7.13x = 0$$
$$x = \frac{3499}{713}$$
$$x \approx 4.91.$$

The solution set is $\{4.91\}$.

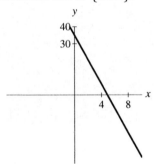

83. a) The midpoint is

$$\left(\frac{0+30}{2}, \frac{20.8 + 25.1}{2}\right) = (15, 22.95).$$

The median of age at first marriage in 1985 was 22.95 years.

b) The distance is

$$\sqrt{(2000 - 1970)^2 + (25.1 - 20.8)^2} \approx 30.3$$

Because of the units, the distance is meaningless.

85. Given $D = 22,800$ lbs, the graph of

$$C = \frac{4B}{\sqrt[3]{22,800}}$$

is given below.

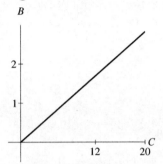

For Island Packet 40, we find

$$C = \frac{4(12 + \frac{11}{12})}{\sqrt[3]{22,800}} \approx 1.8.$$

For Thought

1. False, the slope is $\dfrac{3-2}{3-2} = 1$.

2. False, the slope is $\dfrac{5-1}{-3-(-3)} = \dfrac{4}{0}$

which is undefined.

3. False, slopes of vertical lines are undefined.

4. False, it is a vertical line.   **5.**   True

6. False, $x = 1$ cannot be written in the slope-intercept form.

7. False, the slope is -2.

8. True   **9.** False   **10.** True

1.4 Exercises

1. $\dfrac{5-3}{4+2} = \dfrac{1}{3}$

3. $\dfrac{3+5}{1-3} = -4$

5. $\dfrac{2-2}{5+3} = 0$

7. $\dfrac{1/2 - 1/4}{1/4 - 1/8} = \dfrac{1/4}{1/8} = 2$

9. $\dfrac{3-(-1)}{5-5} = \dfrac{4}{0}$, no slope

11. The slope is $m = \dfrac{4-(-1)}{3-(-1)} = \dfrac{5}{4}$.

Since $y + 1 = \dfrac{5}{4}(x + 1)$, we get

$y = \dfrac{5}{4}x + \dfrac{5}{4} - 1$ or $y = \dfrac{5}{4}x + \dfrac{1}{4}$.

13. The slope is $m = \dfrac{-1-6}{4-(-2)} = -\dfrac{7}{6}$.

Since $y + 1 = -\dfrac{7}{6}(x - 4)$, we obtain

$y = -\dfrac{7}{6}x + \dfrac{14}{3} - 1$ or $y = -\dfrac{7}{6}x + \dfrac{11}{3}$.

15. The slope is $m = \dfrac{5-5}{-3-3} = 0$.

Since $y - 5 = 0(x - 3)$, we get $y = 5$.

17. Since $m = \dfrac{12-(-3)}{4-4} = \dfrac{15}{0}$ is undefined,

the equation of the vertical line is $x = 4$.

19. The slope of the line through $(0, -1)$ and $(3, 1)$

is $m = \dfrac{2}{3}$. Since the y-intercept is $(0, -1)$,

the line is given by $y = \dfrac{2}{3}x - 1$.

21. The slope of the line through $(1, 4)$ and $(-1, 1)$

is $m = \dfrac{5}{2}$. Solving for y in $y - 1 = \dfrac{5}{2}(x + 1)$,

we get $y = \dfrac{5}{2}x + \dfrac{3}{2}$.

23. The slope of the line through $(0, 4)$ and $(2, 0)$

is $m = -2$. Since the y-intercept is $(0, 4)$,
the line is given by $y = -2x + 4$.

25. The slope of the line through $(1, 4)$ and

$(-3, -2)$ is $m = \dfrac{3}{2}$. Solving for y in

$y - 4 = \dfrac{3}{2}(x - 1)$, we get

$$y = \dfrac{3}{2}x + \dfrac{5}{2}.$$

27. $y = \dfrac{3}{5}x - 2$, slope is $\dfrac{3}{5}$, y-intercept is $(0, -2)$

29. Since $y - 3 = 2x - 8$, $y = 2x - 5$.
The slope is 2 and y-intercept is $(0, -5)$.

31. Since $y + 1 = \dfrac{1}{2}x + \dfrac{3}{2}$, $y = \dfrac{1}{2}x + \dfrac{1}{2}$.

The slope is $\dfrac{1}{2}$ and y-intercept is $\left(0, \dfrac{1}{2}\right)$.

33. Since $y = 4$, the slope is $m = 0$ and the
y-intercept is $(0, 4)$.

35. Since $y - 0.4 = 0.03x - 3$, $y = 0.03x - 2.6$.
The slope is 0.03 and y-intercept is $(0, -2.6)$

37. $y = \dfrac{1}{2}x - 2$ goes through the points

$(0, -2), (2, -1)$, and $(4, 0)$.

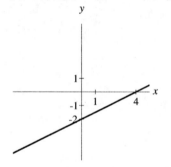

39. $y = -3x + 1$ goes through $(0, 1), (1, -2)$

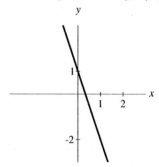

41. $y = -\dfrac{3}{4}x - 1$ goes through $(0, -1), (-4/3, 0)$

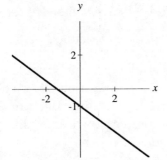

43. $x - y = 3$ goes through $(0, -3), (3, 0)$

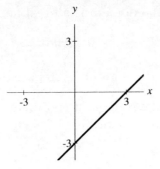

45. $y = 5$ is a horizontal line

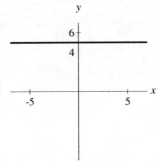

47. Since $m = \dfrac{4}{3}$ and $y - 0 = \dfrac{4}{3}(x - 3)$,

we have $4x - 3y = 12$.

49. Since $m = \dfrac{4}{5}$ and $y - 3 = \dfrac{4}{5}(x - 2)$,

we obtain $5y - 15 = 4x - 8$ and $4x - 5y = -7$.

51. $x = -4$ is a vertical line.

53. 0.5 **55.** -1 **57.** 0

59. Since $y + 2 = 2(x - 1)$, we obtain

$$2x - y = 4.$$

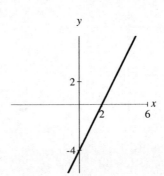

61. Since slope of $y = -3x$ is -3 and
$y - 4 = -3(x - 1)$, we obtain

$$3x + y = 7.$$

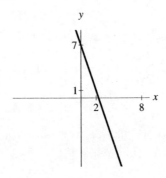

63. Since slope of $y = \dfrac{1}{2}x - \dfrac{3}{2}$ is $\dfrac{1}{2}$ and
$y - 1 = -2(x + 3)$, we find $2x + y = -5$.

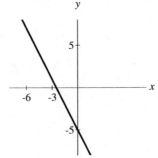

65. Since $x = 4$ is a vertical line, the horizontal line through $(2, 5)$ is $y = 5$.

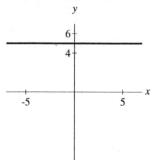

67. The slope is $\dfrac{212 - 32}{100 - 0} = \dfrac{9}{5}$.

Since $F - 32 = \dfrac{9}{5}(C - 0)$, we find

$$F = \frac{9}{5}C + 32.$$

When $C = 150$, $F = \dfrac{9}{5}(150) + 32 = 302^o\text{F}$.

69. A linear function through $(1, 49)$ and $(2, 48)$ is

$$c = 50 - n.$$

With $n = 40$ people in a tour, she would charge $10 each and make $400.

71. Let c and p be the number of computers and printers, respectively. Since

$$60000 = 2000c + 1500p,$$

we have

$$
\begin{aligned}
2000c &= -1500p + 60000 \\
c &= -\frac{3}{4}p + 30.
\end{aligned}
$$

The slope is $-\dfrac{3}{4}$, i.e., if 4 more printers are purchased then 3 fewer computers must be purchased.

For Thought

1. False, since $\{(1, 2), (1, 3)\}$ is not a function.

2. False, since $f(5)$ is not defined.   **3.**   True

4. False, since a student's exam grade is a function of the student's preparation. If two classmates had the same IQ and only one prepared then the one who prepared will most likely achieve a higher grade.

5. False, since $(x + h)^2 = x^2 + 2xh + h^2$

6. False, since the domain is all real numbers.

7. True   **8.**   True   **9.**   True

10. False, since $\left(\dfrac{3}{8}, 8\right)$ and $\left(\dfrac{3}{8}, 5\right)$ are two ordered pairs with the same first coordinate and different second coordinates.

1.5 Exercises

1. Note, $b = 2\pi a$ is equivalent to $a = \dfrac{b}{2\pi}$.

Thus, a is a function of b, and b is a function of a.

3. a is a function of b since a given denomination has a unique length. Since a dollar bill and a five-dollar bill have the same length, then b is not a function of a.

5. Since an item has only one price, b is a function of a. Since two items may have the same price, a is not a function of b.

7. a is not a function of b since it is possible that two different students can obtain the same final exam score but the times spent on studying are different.

b is not a function of a since it is possible that two different students can spend the same time studying but obtain different final exam scores.

9. Since 1 in ≈ 2.54 cm, a is a function of b and b is a function of a.

11. No   **13.**   Yes   **15.**   Yes

17. Function

19. Not a function since 25 has two different second coordinates.

21. Not a function since 3 has two different second coordinates.

23. Function

25. Since the ordered pairs in the graph of $y = 3x - 8$ are $(x, 3x - 8)$, there are no two ordered pairs with the same first coordinate and different second coordinates. We have a function.

27. Since $y = (x + 9)/3$, the ordered pairs are $(x, (x + 9)/3)$. Thus, there are no two ordered pairs with the same first coordinate and different second coordinates. We have a function.

29. Since $y = \pm x$, the ordered pairs are $(x, \pm x)$. Thus, there are two ordered pairs with the same first coordinate and different second coordinates. We do not have a function.

31. Since $y = x^2$, the ordered pairs are (x, x^2). Thus, there are no two ordered pairs with the same first coordinate and different second coordinates. We have a function.

33. Since $y = |x| - 2$, the ordered pairs are $(x, |x| - 2)$. Thus, there are no two ordered pairs with the same first coordinate and different second coordinates. We have a function.

35. Since $(2, 1)$ and $(2, -1)$ are two ordered pairs with the same first coordinate and different second coordinates, the equation does not define a function.

37. domain $\{-3, 4, 5\}$, range $\{1, 2, 6\}$

39. Domain $(-\infty, \infty)$, range $\{4\}$

41. Domain $(-\infty, \infty)$;
since $|x| \geq 0$, the range of $y = |x| + 5$ is $[5, \infty)$

43. since $|y| - 3 \geq -3$, the domain of $x = |y| - 3$ is $[-3, \infty)$; range $(-\infty, \infty)$

45. Since $\sqrt{x - 4}$ is a real number whenever $x \geq 4$, the domain of $y = \sqrt{x - 4}$ is $[4, \infty)$.

Since $y = \sqrt{x - 4} \geq 0$ for $x \geq 4$, the range is $[0, \infty)$.

47. Since $x = -y^2 \leq 0$, the domain of $x = -y^2$ is $(-\infty, 0]$; range is $(-\infty, \infty)$;

49. 6

51. $g(2) = 3(2) + 5 = 11$

53. Since $(3, 8)$ is the ordered pair, one obtains $f(3) = 8$. Then $x = 3$.

55. Solving for $3x + 5 = 26$, we find $x = 7$.

57. $f(4) + g(4) = 5 + 17 = 22$

59. $3a^2 - a$

61. $4(a + 2) - 2 = 4a + 6$

63. $3(x^2 + 2x + 1) - (x + 1) = 3x^2 + 5x + 2$

65. $4(x + h) - 2 = 4x + 4h - 2$

67. $(3x^2 + 6xh + 3h^2 - x - h) - 3x^2 + x = 3h^2 + 6xh - h$

69. The average rate of change is

$$\frac{4,000 - 16,000}{5} = -\$2,400 \text{ per year.}$$

71. The average rate of change on $[0, 2]$ is
$$\frac{h(2) - h(0)}{2 - 0} = \frac{0 - 64}{2 - 0} = -32 \text{ ft/sec.}$$
The average rate of change on $[1, 2]$ is
$$\frac{h(2) - h(1)}{2 - 1} = \frac{0 - 48}{2 - 1} = -48 \text{ ft/sec.}$$
The average rate of change on $[1.9, 2]$ is
$$\frac{h(2) - h(1.9)}{2 - 1.9} = \frac{0 - 6.24}{0.1} = -62.4 \text{ ft/sec.}$$
The average rate of change on $[1.99, 2]$ is
$$\frac{h(2) - h(1.99)}{2 - 1.99} = \frac{0 - 0.6384}{0.01} = -63.84 \text{ ft/sec.}$$
The average rate of change on $[1.999, 2]$ is
$$\frac{h(2) - h(1.999)}{2 - 1.999} = \frac{0 - 0.063984}{0.001} = -63.984$$
ft/sec.

73.

$$\frac{f(x + h) - f(x)}{h} = \frac{4(x + h) - 4x}{h}$$
$$= \frac{4h}{h}$$
$$= 4$$

75.

$$\frac{f(x+h)-f(x)}{h} = \frac{3(x+h)+5-3x-5}{h}$$
$$= \frac{3h}{h}$$
$$= 3$$

77. Let $g(x) = x^2 + x$. Then we obtain

$$\frac{g(x+h)-g(x)}{h} =$$
$$\frac{(x+h)^2 + (x+h) - x^2 - x}{h} =$$
$$\frac{2xh + h^2 + h}{h} =$$
$$2x + h + 1.$$

79. Difference quotient is

$$= \frac{-(x+h)^2 + (x+h) - 2 + x^2 - x + 2}{h}$$
$$= \frac{-2xh - h^2 + h}{h}$$
$$= -2x - h + 1$$

81. Difference quotient is

$$= \frac{3\sqrt{x+h} - 3\sqrt{x}}{h} \cdot \frac{3\sqrt{x+h} + 3\sqrt{x}}{3\sqrt{x+h} + 3\sqrt{x}}$$
$$= \frac{9(x+h) - 9x}{h(3\sqrt{x+h} + 3\sqrt{x})}$$
$$= \frac{9h}{h(3\sqrt{x+h} + 3\sqrt{x})}$$
$$= \frac{3}{\sqrt{x+h} + \sqrt{x}}$$

83. Difference quotient is

$$= \frac{\sqrt{x+h+2} - \sqrt{x+2}}{h} \cdot \frac{\sqrt{x+h+2} + \sqrt{x+2}}{\sqrt{x+h+2} + \sqrt{x+2}}$$
$$= \frac{(x+h+2) - (x+2)}{h(\sqrt{x+h+2} + \sqrt{x+2})}$$
$$= \frac{h}{h(\sqrt{x+h+2} + \sqrt{x+2})}$$
$$= \frac{1}{\sqrt{x+h+2} + \sqrt{x+2}}$$

85. Difference quotient is

$$= \frac{\frac{1}{x+h} - \frac{1}{x}}{h} \cdot \frac{x(x+h)}{x(x+h)}$$
$$= \frac{x - (x+h)}{xh(x+h)}$$
$$= \frac{-h}{xh(x+h)}$$
$$= \frac{-1}{x(x+h)}$$

87. Difference quotient is

$$= \frac{\frac{3}{x+h+2} - \frac{3}{x+2}}{h} \cdot \frac{(x+h+2)(x+2)}{(x+h+2)(x+2)}$$
$$= \frac{3(x+2) - 3(x+h+2)}{h(x+h+2)(x+2)}$$
$$= \frac{-3h}{h(x+h+2)(x+2)}$$
$$= \frac{-3}{(x+h+2)(x+2)}$$

89. a) $A = s^2$ **b)** $s = \sqrt{A}$ **c)** $s = \frac{\sqrt{2}d}{2}$
   d) $d = \sqrt{2}s$ **e)** $P = 4s$ **f)** $s = P/4$
   g) $A = P^2/16$ **h)** $d = \sqrt{2A}$

91. a) When $d = 100$ ft, atmospheric pressure is

$$A(100) = .03(100) + 1 = 4 \text{ atm.}$$

   b) When $A = 4.9$ atm, the depth is found by solving $4.9 = 0.03d + 1$; the depth is

$$d = \frac{3.9}{0.03} = 130 \text{ ft.}$$

93. Let a be the radius of each circle. Note, triangle $\triangle ABC$ is an equilateral triangle with side $2a$ and height $\sqrt{3}a$.

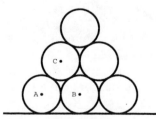

Thus, the height of the circle centered at C from the horizontal line is $\sqrt{3}a + 2a$. Hence, by using a similar reasoning, we obtain that height of the highest circle from the line is

$$2\sqrt{3}a + 2a$$

or equivalently $(2\sqrt{3} + 2)a$.

For Thought

1. True, since the graph is a parabola opening down with vertex at the origin.

2. False, the graph is decreasing.

3. True

4. True, since $f(-4.5) = [-1.5] = -2$.

5. False, since the range is $\{\pm 1\}$.

6. True   7.   True   8.   True

9. False, since the range is the interval $[0, 4]$.

10. True

1.6 Exercises

1. Function $y = 2x$ includes the points $(0,0), (1,2)$, domain and range are both $(-\infty, \infty)$

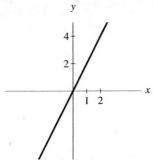

3. Function $x - y = 0$ includes the points $(-1, -1)$, $(0, 0), (1, 1)$, domain and range are both $(-\infty, \infty)$

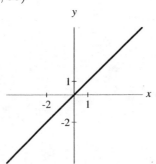

5. Function $y = 5$ includes the points $(0, 5)$, $(\pm 2, 5)$, domain is $(-\infty, \infty)$, range is $\{5\}$

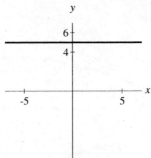

7. Function $y = 2x^2$ includes the points $(0, 0)$, $(\pm 1, 2)$, domain is $(-\infty, \infty)$, range is $[0, \infty)$

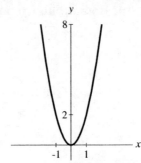

9. Function $y = 1 - x^2$ includes the points $(0, 1)$, $(\pm 1, 0)$, domain is $(-\infty, \infty)$, range is $(-\infty, 1]$

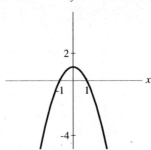

11. Function $y = 1 + \sqrt{x}$ includes the points $(0, 1)$, $(1, 2), (4, 3)$, domain is $[0, \infty)$, range is $[1, \infty)$

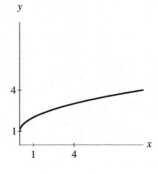

13. $x = y^2 + 1$ is not a function and includes the points $(1,0), (2, \pm 1)$, domain is $[1, \infty)$, range is $(-\infty, \infty)$

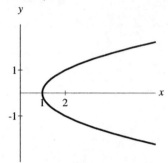

15. Function $x = \sqrt{y}$ goes through $(0,0), (2,4), (3,9)$, domain and range is $[0, \infty)$

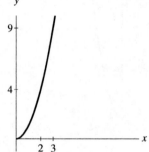

17. Function $y = \sqrt[3]{x} + 1$ goes through $(-1, 0), (1, 2), (8, 3)$, domain $(-\infty, \infty)$, and range $(-\infty, \infty)$

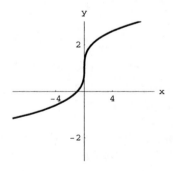

19. Function, $x = \sqrt[3]{y}$ goes through $(0,0), (1,1), (2,8)$, domain $(-\infty, \infty)$, and range $(-\infty, \infty)$

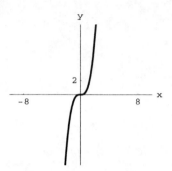

21. Not a function, $y^2 = 1 - x^2$ goes through $(1,0), (0,1), (-1,0)$, domain $[-1, 1]$, and range $[-1, 1]$

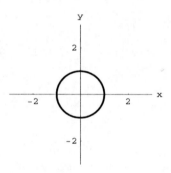

23. Function, $y = \sqrt{1 - x^2}$ goes through $(\pm 1, 0), (0, 1)$, domain $[-1, 1]$, and range $[0, 1]$

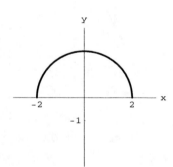

25. Function $y = x^3 + 1$ includes the points $(0, 1)$, $(1, 2), (2, 9)$, domain and range are both $(-\infty, \infty)$

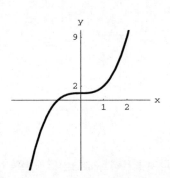

27. Function $y = 2|x|$ includes the points $(0, 0)$, $(\pm 1, 2)$, domain is $(-\infty, \infty)$, range is $[0, \infty)$

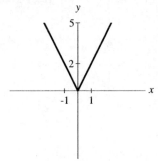

29. Function $y = -|x|$ includes the points $(0, 0)$, $(\pm 1, -1)$, domain is $(-\infty, \infty)$, range is $(-\infty, 0]$

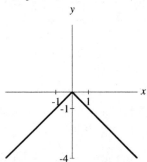

31. Not a function, graph of $x = |y|$ includes the points $(0, 0), (2, 2), (2, -2)$, domain is $[0, \infty)$, range is $(-\infty, \infty)$

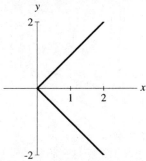

33. Domain is $(-\infty, \infty)$, range is $\{\pm 2\}$, some points are $(-3, -2)$, $(1, -2)$

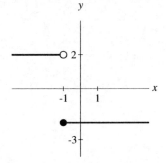

35. Domain is $(-\infty, \infty)$, range is $(-\infty, -2] \cup (2, \infty)$, some points are $(2, 3)$, $(1, -2)$

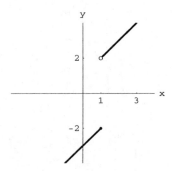

37. Domain is $[-2, \infty)$, range is $(-\infty, 2]$, some points are $(2, 2)$, $(-2, 0)$, $(3, 1)$

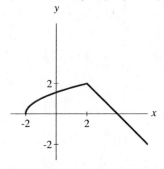

39. Domain is $(-\infty, \infty)$, range is $[0, \infty)$, some points are $(-1, 1)$, $(-4, 2)$, $(4, 2)$

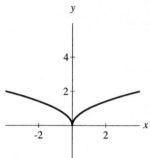

41. Domain is $(-\infty, \infty)$, range is $(-\infty, \infty)$, some points are $(-2, 4)$, $(1, -1)$

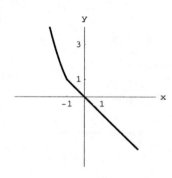

43. Domain is $(-\infty, \infty)$, range is the set of integers, some points are $(0, 1)$, $(1, 2)$, $(1.5, 2)$

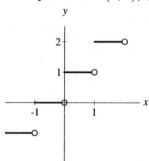

45. Domain $[0, 4)$, range is $\{2, 3, 4, 5\}$, some points are $(0, 2)$, $(1, 3)$, $(1.5, 3)$

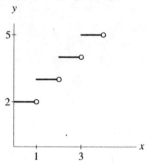

47. a. Domain and range are both $(-\infty, \infty)$, decreasing on $(-\infty, \infty)$

b. Domain is $(-\infty, \infty)$, range is $(-\infty, 4]$ increasing on $(-\infty, 0)$, decreasing on $(0, \infty)$

49. a. Domain is $[-2, 6]$, range is $[3, 7]$ increasing on $(-2, 2)$, decreasing on $(2, 6)$

b. Domain $(-\infty, 2]$, range $(-\infty, 3]$, increasing on $(-\infty, -2)$, constant on $(-2, 2)$

51. a. Domain is $(-\infty, \infty)$, range is $[0, \infty)$ increasing on $(0, \infty)$, decreasing on $(-\infty, 0)$

b. Domain and range are both $(-\infty, \infty)$ increasing on $(-2, -2/3)$, decreasing on $(-\infty, -2)$ and $(-2/3, \infty)$

53. a. Domain and range are both $(-\infty, \infty)$, increasing on $(-\infty, \infty)$

b. Domain is $[-2, 5]$, range is $[1, 4]$ increasing on $(1, 2)$, decreasing on $(-2, 1)$, constant on $(2, 5)$

55. Domain and range are both $(-\infty, \infty)$ increasing on $(-\infty, \infty)$, some points are $(0, 1)$, $(1, 3)$

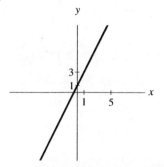

57. Domain is $(-\infty, \infty)$, range is $[0, \infty)$, increasing on $(1, \infty)$, decreasing on $(-\infty, 1)$, some points are $(0, 1)$, $(1, 0)$

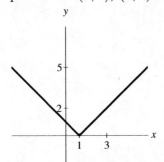

59. Domain is $(-\infty, 0) \cup (0, \infty)$, range is $\{\pm 1\}$, constant on $(-\infty, 0)$ and $(0, \infty)$, some points are $(1, 1)$, $(-1, -1)$

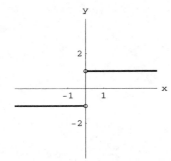

61. Domain is $[-3, 3]$, range is $[0, 3]$, increasing on $(-3, 0)$, decreasing on $(0, 3)$, some points are $(\pm 3, 0)$, $(0, 3)$

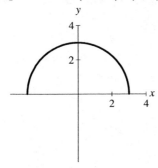

63. Domain and range are both $(-\infty, \infty)$, increasing on $(-\infty, 3)$ and $(3, \infty)$, some points are $(4, 5)$, $(0, 2)$

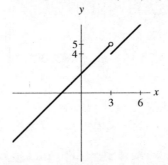

65. Domain is $(-\infty, \infty)$, range is $(-\infty, 2]$, increasing on $(-\infty, -2)$ and $(-2, 0)$, decreasing on $(0, 2)$ and $(2, \infty)$, some points are $(-3, 0)$, $(0, 2)$, $(4, -1)$

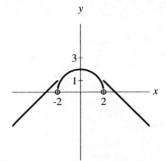

67. $f(x) = \begin{cases} 2 & \text{for} \quad x > -1 \\ -1 & \text{for} \quad x \le -1 \end{cases}$

69. The line joining $(-1, 1)$ and $(-3, 3)$ is $y = -x$, and the line joining $(-1, -2)$ and $(3, 2)$ is $y = x - 1$. The piecewise function is

$$f(x) = \begin{cases} x - 1 & \text{for} \quad x \ge -1 \\ -x & \text{for} \quad x < -1. \end{cases}$$

71. The line joining $(0, -2)$ and $(2, 2)$ is $y = 2x - 2$, and the line joining $(0, -2)$ and $(-3, 1)$ is $y = -x - 2$. The piecewise function is

$$f(x) = \begin{cases} 2x - 2 & \text{for} \quad x \ge 0 \\ -x - 2 & \text{for} \quad x < 0. \end{cases}$$

73. increasing on the interval $(0.83, \infty)$,

decreasing on $(-\infty, 0.83)$

75. increasing on $(-\infty, -1)$ and $(1, \infty)$,
decreasing on $(-1, 1)$

77. increasing on $(-1.73, 0)$ and $(1.73, \infty)$,
decreasing on $(-\infty, -1.73)$ and $(0, 1.73)$

79. increasing on $(30, 50)$, and $(70, \infty)$,
decreasing on $(-\infty, 30)$ and $(50, 70)$

81. In 1988, there were $M(18) = 565$ million cars.
In 2010, it is projected that there will be
$M(40) = 800$ million cars.
The average rate of change from 1984 to 1994
is $\dfrac{M(24) - M(14)}{10} = 14.5$ million cars per
year.

83. The cost is over \$235 for t in $[5, \infty)$.

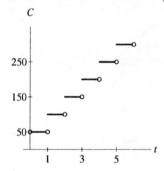

For Thought

1. False, it is a reflection in the y-axis.

2. True   **3.**   False, rather it is a left translation.

4. True   **5.**   True

6. False, the down shift should come after the
reflection.   **7.**   True

8. False, since their domains are different.

9. True   **10.**   True

1.7 Exercises

1. $f(x) = |x|, g(x) = |x| - 4$

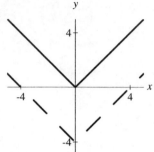

3. $f(x) = x, g(x) = x + 3$

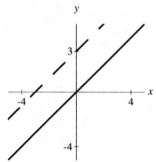

5. $y = x^2, y = (x - 3)^2$

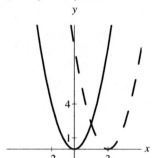

7. $y = \sqrt{x}, y = \sqrt{x + 9}$

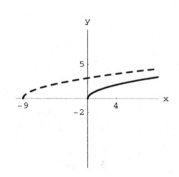

9. $f(x) = \sqrt{x}$, $g(x) = -\sqrt{x}$

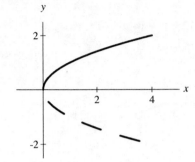

11. $y = \sqrt{x}, y = 3\sqrt{x}$

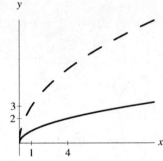

13. $y = x^2, y = \frac{1}{4}x^2$

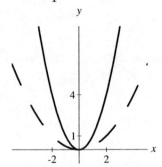

15. $y = \sqrt{4 - x^2}, y = -\sqrt{4 - x^2}$

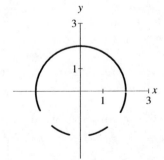

17. g   **19.**   b

21. c   **23.**   f

25. $y = \sqrt{x} + 2$

27. $y = (x - 5)^2$

29. $y = (x - 10)^2 + 4$

31. $y = -(3\sqrt{x} + 5)$ or $y = -3\sqrt{x} - 5$

33. $y = -3|x - 7| + 9$

35. $y = (x - 1)^2 + 2$; right by 1, up by 2, domain $(-\infty, \infty)$, range $[2, \infty)$

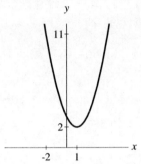

37. $y = |x - 1| + 3$; right by 1, up by 3 domain $(-\infty, \infty)$, range $[3, \infty)$

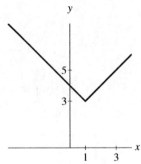

39. $y = 3x - 40$, domain and range are both $(-\infty, \infty)$

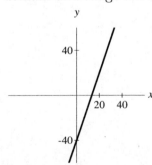

41. $y = \dfrac{1}{2}x - 20$,

domain and range are both $(-\infty, \infty)$

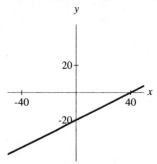

43. $y = -\dfrac{1}{2}|x| + 40$, shrink by 1/2,

reflect about x-axis, up by 40,
domain $(-\infty, \infty)$, range $(-\infty, 40]$

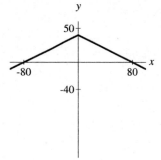

45. $y = -\dfrac{1}{2}|x + 4|$, left by 4,

reflect about x-axis, shrink by 1/2,
domain $(-\infty, \infty)$, range $(-\infty, 0]$

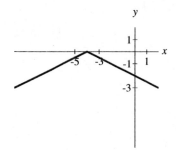

47. $y = -\sqrt{x - 3} + 1$, right by 3,
reflect about x-axis, up by 1,
domain $[3, \infty)$, range $(-\infty, 1]$

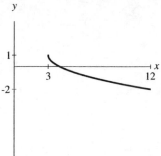

49. $y = -2\sqrt{x + 3} + 2$, left by 3, stretch by 2,
reflect about x-axis, up by 2,
domain $[-3, \infty)$, range $(-\infty, 2]$

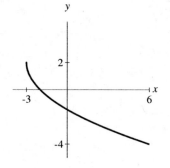

51. Symmetric about y-axis, even function
since $f(-x) = f(x)$

53. No symmetry, neither even nor odd
since $f(-x) \neq f(x)$ and $f(-x) \neq -f(x)$

55. Symmetric about $x = -3$, neither even nor
odd since $f(-x) \neq f(x)$ and $f(-x) \neq -f(x)$

57. Symmetry about $x = 2$, not an even or odd
function since $f(-x) \neq f(x)$ and
$f(-x) \neq -f(x)$

59. Symmetric about the origin, odd function
since $f(-x) = -f(x)$

61. No symmetry, not an even or odd function
since $f(-x) \neq f(x)$ and $f(-x) \neq -f(x)$

63. No symmetry, not an even or odd function
since $f(-x) \neq f(x)$ and $f(-x) \neq -f(x)$

65. Symmetric about the y-axis, even function
since $f(-x) = f(x)$

67. e   **69.**   g

71. b   **73.**   c

75. $(-\infty, -1] \cup [1, \infty)$

77. $(-\infty, -1) \cup (5, \infty)$

79. Using the graph of $y = (x-1)^2 - 9$, we find that the solution is $(-2, 4)$.

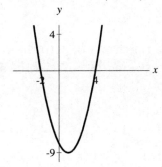

81. From the graph of $y = 5 - \sqrt{x}$, we find that the solution set is $[0, 25]$.

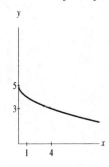

83. Note, the points of intersection of $y = 3$ and $y = (x-2)^2$ are $(2 \pm \sqrt{3}, 3)$. The solution set of $(x-2)^2 > 3$ is

$$\left(-\infty, 2 - \sqrt{3}\right) \cup \left(2 + \sqrt{3}, \infty\right).$$

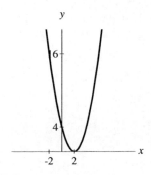

85. From the graph of $y = \sqrt{25 - x^2}$, we conclude that the solution is $(-5, 5)$.

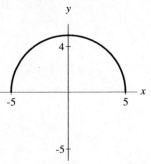

87. From the graph of $y = \sqrt{3}x^2 + \pi x - 9$,

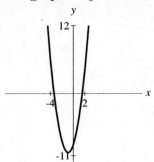

we observe that the solution set of $\sqrt{3}x^2 + \pi x - 9 < 0$ is $(-3.36, 1.55)$.

89. a. Stretch the graph of f by a factor of 2.

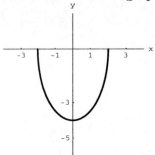

b. Reflect the graph of f about the x-axis.

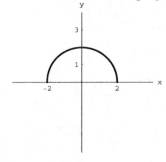

c. Translate the graph of f to the left by 1-unit.

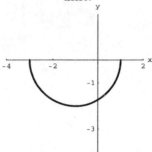

d. Translate the graph of f to the right by 3-units.

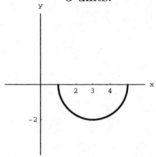

e. Stretch the graph of f by a factor of 3 and reflect about the x-axis.

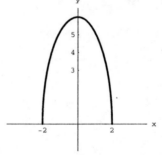

f. Translate the graph of f to the left by 2-units and down by 1-unit.

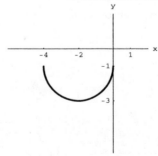

g. Translate the graph of f to the right by 1-unit and up by 3-units.

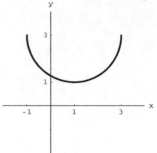

h. Translate the graph of f to the right by 2-units, stretch by a factor of 3, and up by 1-unit.

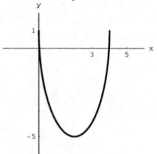

91. $N(x) = x + 2000$

For Thought

1. False, since $f + g$ has an empty domain.

2. True   **3.**   True   **4.**   True

5. True, since $A = P^2/16$.   **6.**   True

7. False, since $(f \circ g)(x) = \sqrt{x-2}$ **8.**   True

9. False, since $(h \circ g)(x) = x^2 - 9$.

10. True, since x belongs to the domain if $\sqrt{x-2}$ is a real number, i.e., if $x \geq 2$.

1.8 Exercises

1. $-1 + 2 = 1$

3. $-5 - 6 = -11$

5. $(-4) \cdot 2 = -8$

7. $1/12$

9. $(a - 3) + (a^2 - a) = a^2 - 3$

11. $(a-3)(a^2-a) = a^3 - 4a^2 + 3a$

13. $f+g = \{(-3, 1+2), (2, 0+6)\} = \{(-3, 3), (2, 6)\}$, domain $\{-3, 2\}$

15. $f-g = \{(-3, 1-2), (2, 0-6)\} = \{(-3, -1), (2, -6)\}$, domain $\{-3, 2\}$

17. $f \cdot g = \{(-3, 1 \cdot 2), (2, 0 \cdot 6) = \{(-3, 2), (2, 0)\}$, domain $\{-3, 2\}$

19. $g/f = \{(-3, 2/1)\} = \{(-3, 2)\}$, domain $\{-3\}$

21. $(f+g)(x) = \sqrt{x} + x - 4$, domain is $[0, \infty)$

23. $(f-h)(x) = \sqrt{x} - \dfrac{1}{x-2}$, domain is $[0, 2) \cup (2, \infty)$

25. $(g \cdot h)(x) = \dfrac{x-4}{x-2}$, domain is $(-\infty, 2) \cup (2, \infty)$

27. $\left(\dfrac{g}{f}\right)(x) = \dfrac{x-4}{\sqrt{x}}$, domain is $(0, \infty)$

29. $\{(-3, 0), (1, 0), (4, 4)\}$

31. $\{(1, 4)\}$

33. $\{(-3, 4), (1, 4)\}$

35. $f(2) = 5$

37. $f(2) = 5$

39. $f(20.2721) = 59.8163$

41. $(g \circ h \circ f)(2) = (g \circ h)(5) = g(2) = 5$

43. $(f \circ g \circ h)(2) = (f \circ g)(1) = f(2) = 5$

45. $(f \circ h)(a) = f\left(\dfrac{a+1}{3}\right) = 3\left(\dfrac{a+1}{3}\right) - 1 = (a+1) - 1 = a$

47. $(f \circ g)(t) = f(t^2 + 1) = 3(t^2 + 1) - 1 = 3t^2 + 2$

49. $(f \circ g)(x) = \sqrt{x} - 2$, domain $[0, \infty)$

51. $(f \circ h)(x) = \dfrac{1}{x} - 2$, domain $(-\infty, 0) \cup (0, \infty)$

53. $(h \circ g)(x) = \dfrac{1}{\sqrt{x}}$, domain $(0, \infty)$

55. $(f \circ f)(x) = (x-2) - 2 = x - 4$, domain $(-\infty, \infty)$

57. $(h \circ g \circ f)(x) = h(\sqrt{x-2}) = \dfrac{1}{\sqrt{x-2}}$, domain $(2, \infty)$

59. $(h \circ f \circ g)(x) = h(\sqrt{x} - 2) = \dfrac{1}{\sqrt{x} - 2}$, domain $(0, 4) \cup (4, \infty)$

61. $F = g \circ h$

63. $H = h \circ g$

65. $N = h \circ g \circ f$

67. $P = g \circ f \circ g$

69. $S = g \circ g$

71. $y = 2(3x+1) - 3 = 6x - 1$

73. $y = (x^2 + 6x + 9) - 2 = x^2 + 6x + 7$

75. $y = 3 \cdot \dfrac{x+1}{3} - 1 = x + 1 - 1 = x$

77. Domain $[-1, \infty)$, range $[-7, \infty)$

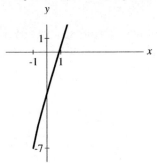

79. Domain $[1, \infty)$, range $[0, \infty)$

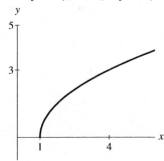

81. Domain $[0, \infty)$, range $[4, \infty)$

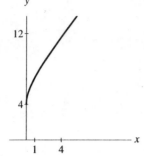

83. $P(x) = 68x - (40x + 200) = 28x - 200$.

Since $200/28 \approx 7.1$, the profit is positive when the number of trimmers satisfies $x \geq 8$.

85. $A = d^2/2$

87. Total cost is $T(x) = 1.05(1.20x) = 1.26x$.

89. Note, $D = \dfrac{d/2240}{x} = \dfrac{d/2240}{L^3/100^3} = \dfrac{100^3 d}{2240 L^3}$

$= \dfrac{100^3 (26000)}{2240 L^3} = \dfrac{100^4 (26)}{224 L^3} = \dfrac{100^4 (13)}{112 L^3}$.

Expressing D as a function of L, we

write $D = \dfrac{(13)100^4}{112 L^3}$ or $D = \dfrac{1.16 \times 10^7}{L^3}$.

91. The area of a semicircle with radius $s/2$ is $(1/2)\pi(s/2)^2 = \pi s^2/8$. The area of the square is s^2. The area of the window is

$$W = s^2 + \frac{\pi s^2}{8} = \frac{(8 + \pi)s^2}{8}.$$

For Thought

1. False, since the inverse function is $\{(3, 2), (5, 5)\}$.

2. False, since it is not one-to-one.

3. False, $g^{-1}(x)$ does not exist since g is not one-to-one.

4. True

5. False, a function that fails the horizontal line test has no inverse.

6. False, since it fails the horizontal line test.

7. False, since $f^{-1}(x) = \left(\dfrac{x}{3}\right)^2 + 2$ where $x \geq 0$.

8. False, $f^{-1}(x)$ does not exist since f is not one-to-one.

9. False, since $y = |x|$ is V-shaped and the horizontal line test fails.

10. True

1.9 Exercises

1. Yes, since all second coordinates are distinct.

3. No, since there are repeated second coordinates such as $(-1, 1)$ and $(1, 1)$.

5. No, since there are repeated second coordinates such as $(1, 99)$ and $(5, 99)$.

7. Not one-to-one

9. One-to-one

11. Not one-to-one

13. One-to-one; since the graph of $y = 2x - 3$ shows $y = 2x - 3$ is an increasing function, the Horizontal Line Test implies $y = 2x - 3$ is one-to-one.

15. One-to-one; for if $q(x_1) = q(x_2)$ then

$$
\begin{aligned}
\frac{1 - x_1}{x_1 - 5} &= \frac{1 - x_2}{x_2 - 5} \\
(1 - x_1)(x_2 - 5) &= (1 - x_2)(x_1 - 5) \\
x_2 - 5 - x_1 x_2 + 5x_1 &= x_1 - 5 - x_2 x_1 + 5x_2 \\
x_2 + 5x_1 &= x_1 + 5x_2 \\
4(x_1 - x_2) &= 0 \\
x_1 - x_2 &= 0.
\end{aligned}
$$

Thus, if $q(x_1) = q(x_2)$ then $x_1 = x_2$. Hence, q is one-to-one.

17. Not one-to-one for $p(-2) = p(0) = 1$.

19. Not one-to-one for $w(1) = w(-1) = 4$.

21. One-to-one; for if $k(x_1) = k(x_2)$ then

$$
\begin{aligned}
\sqrt[3]{x_1 + 9} &= \sqrt[3]{x_2 + 9} \\
(\sqrt[3]{x_1 + 9})^3 &= (\sqrt[3]{x_2 + 9})^3 \\
x_1 + 9 &= x_2 + 9 \\
x_1 &= x_2.
\end{aligned}
$$

Thus, if $k(x_1) = k(x_2)$ then $x_1 = x_2$. Hence, k is one-to-one.

23. Invertible, $\{(3,9),(2,2)\}$

25. Not invertible

27. Invertible, $\{(3,3),(2,2),(4,4),(7,7)\}$

29. Not invertible

31. Not invertible, there can be two different items with the same price.

33. Invertible, since the playing time is a function of the length of the VCR tape.

35. Invertible, assuming that cost is simply a multiple of the number of days. If cost includes extra charges, then the function may not be invertible.

37. $f^{-1} = \{(1,2),(5,3)\}$, $f^{-1}(5) = 3$, $(f^{-1} \circ f)(2) = 2$

39. $f^{-1} = \{(-3,-3),(5,0),(-7,2)\}$, $f^{-1}(5) = 0$, $(f^{-1} \circ f)(2) = 2$

41. Not invertible since it fails the Horizontal Line Test.

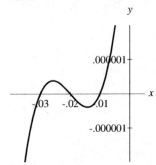

43. Not invertible since it fails the Horizontal Line Test.

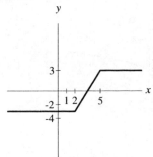

45. Interchange x and y then solve for y.

$$\begin{aligned} x &= 3y - 7 \\ \frac{x+7}{3} &= y \\ \frac{x+7}{3} &= f^{-1}(x) \end{aligned}$$

47. Interchange x and y then solve for y.

$$\begin{aligned} x &= 2 + \sqrt{y-3} \quad \text{for } x \geq 2 \\ (x-2)^2 &= y - 3 \quad \text{for } x \geq 2 \\ f^{-1}(x) &= (x-2)^2 + 3 \quad \text{for } x \geq 2 \end{aligned}$$

49. Interchange x and y then solve for y.

$$\begin{aligned} x &= -y - 9 \\ y &= -x - 9 \\ f^{-1}(x) &= -x - 9 \end{aligned}$$

51. Interchange x and y then solve for y.

$$\begin{aligned} x &= \frac{y+3}{y-5} \\ xy - 5x &= y + 3 \\ xy - y &= 5x + 3 \\ y(x-1) &= 5x + 3 \\ f^{-1}(x) &= \frac{5x+3}{x-1} \end{aligned}$$

53. Interchange x and y then solve for y.

$$\begin{aligned} x &= -\frac{1}{y} \\ xy &= -1 \\ f^{-1}(x) &= -\frac{1}{x} \end{aligned}$$

55. Interchange x and y then solve for y.

$$\begin{aligned} x &= \sqrt[3]{y-9} + 5 \\ x - 5 &= \sqrt[3]{y-9} \\ (x-5)^3 &= y - 9 \\ f^{-1}(x) &= (x-5)^3 + 9 \end{aligned}$$

57. Interchange x and y then solve for y.

$$\begin{aligned} x &= (y-2)^2 \quad x \geq 0 \\ \sqrt{x} &= y - 2 \\ f^{-1}(x) &= \sqrt{x} + 2 \end{aligned}$$

59. Since

$$(g \circ f)(x) = 0.25(4x + 4) - 1 = x$$

and

$$(f \circ g)(x) = 4(0.25x - 1) + 4 = x,$$

we obtain that g and f are inverse functions of each other.

61. Since

$$(f \circ g)(x) = \left(\sqrt{x - 1}\right)^2 + 1 = x$$

and

$$(g \circ f)(x) = \sqrt{x^2 + 1 - 1} = \sqrt{x^2} = |x|$$

we find that g and f are not inverse functions of each other.

63. We find

$$\begin{aligned}(f \circ g)(x) &= \frac{1}{1/(x - 3)} + 3 \\ &= x - 3 + 3 \\ (f \circ g)(x) &= x\end{aligned}$$

and

$$\begin{aligned}(g \circ f)(x) &= \frac{1}{\left(\frac{1}{x} + 3\right) - 3} \\ &= \frac{1}{1/x} \\ (g \circ f)(x) &= x.\end{aligned}$$

Then g and f are inverse functions of each other.

65. We obtain

$$\begin{aligned}(f \circ g)(x) &= \sqrt[3]{\frac{5x^3 + 2 - 2}{5}} \\ &= \sqrt[3]{\frac{5x^3}{5}} \\ &= \sqrt[3]{x^3} \\ (f \circ g)(x) &= x\end{aligned}$$

and

$$\begin{aligned}(g \circ f)(x) &= 5\left(\sqrt[3]{\frac{x - 2}{5}}\right)^3 + 2 \\ &= 5\left(\frac{x - 2}{5}\right) + 2 \\ &= (x - 2) + 2 \\ (g \circ f)(x) &= x.\end{aligned}$$

Thus, g and f are inverse functions of each other.

67. y_1 and y_2 are inverse functions of each other and $y_3 = y_2 \circ y_1$.

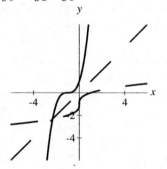

69. No, since they fail the Horizontal Line Test.

71. Yes, since the graphs are symmetric about the line $y = x$.

73. Graph of f^{-1}

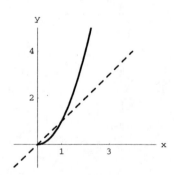

75. Graph of f^{-1}

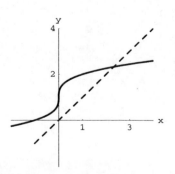

77. $f^{-1}(x) = \dfrac{x-2}{3}$

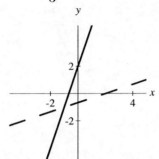

79. $f^{-1}(x) = \sqrt{x+4}$

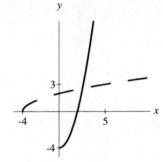

81. $f^{-1}(x) = \sqrt[3]{x}$

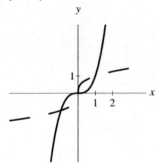

83. $f^{-1}(x) = (x+3)^2$ for $x \geq -3$

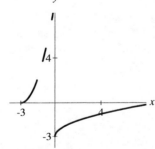

85. a) $f^{-1}(x) = x/5$

b) $f^{-1}(x) = x + 88$

c) $f^{-1}(x) = (x+7)/3$

d) $f^{-1}(x) = \dfrac{x-4}{-3}$

e) $f^{-1}(x) = 2(x+9) = 2x + 18$

f) $f^{-1}(x) = -x$

g) $f^{-1}(x) = (x+9)^3$

h) $f^{-1}(x) = \sqrt[3]{\dfrac{x+7}{3}}$

87. $C = 1.08P$ expresses the total cost as a function of the purchase price; and $P = C/1.08$ is the purchase price as a function of the total cost.

89. The graph of t as a function of r satisfies the Horizontal Line Test and is invertible. Solving for r we find,

$$
\begin{aligned}
t - 7.89 &= -0.39r \\
r &= \frac{t - 7.89}{-0.39}
\end{aligned}
$$

and the inverse function is $r = \dfrac{7.89 - t}{0.39}$.

If $t = 5.55$ min., then $r = \dfrac{7.89 - 5.55}{0.39} = 6$ rowers.

91. Solving for w, we obtain

$$
\begin{aligned}
1.496w &= V^2 \\
w &= \frac{V^2}{1.496}
\end{aligned}
$$

and the inverse function is $w = \dfrac{V^2}{1.496}$. If $V = 115$ ft./sec., then $w = \dfrac{115^2}{1.496} \approx 8,840$ lb.

93. a) Let $V = \$28,000$. The depreciation rate is

$$
r = 1 - \left(\frac{28,000}{50,000}\right)^{1/5} \approx 0.109
$$

or $r \approx 10.9\%$.

b) Writing V as a function of r we find

$$
\begin{aligned}
1 - r &= \left(\frac{V}{50,000}\right)^{1/5} \\
(1 - r)^5 &= \frac{V}{50,000}
\end{aligned}
$$

and $V = 50,000(1 - r)^5$.

Chapter 1 Review Exercises

1. False, since $\sqrt{2}$ is an irrational number.

3. False, since -1 is a negative number.

5. False, since terminating decimal numbers are rational numbers.

7. False, since $\{1, 2, 3, ...\}$ is the set of natural numbers.

9. False, since $\dfrac{1}{3} = 0.3333...$

11. False, since the additive inverse of 0.5 is -0.5

13. Solve an equivalent statement

$$3q - 4 = 2 \quad \text{or} \quad 3q - 4 = -2$$
$$3q = 6 \quad \text{or} \quad 3q = 2.$$

The solution set is $\{2/3, 2\}$.

15. We obtain

$$|2h - 3| = 0$$
$$2h - 3 = 0$$
$$h = \frac{3}{2}.$$

The solution set is $\left\{\dfrac{3}{2}\right\}$.

17. No solution since absolute values are nonnegative.

19. The solution set of $x > 3$ is the interval $(3, \infty)$ and the graph is

21. The solution set of $8 > 2x$ is the interval $(-\infty, 4)$ and the graph is

23. Since $-\dfrac{7}{3} > \dfrac{1}{2}x$, the solution set is

$(-\infty, -14/3)$ and the graph is

25. After multiplying the inequality by 2 we have

$$-4 < x - 3 \le 10$$
$$-1 < x \le 13.$$

The solution set is the interval $(-1, 13]$ and the graph is

27. The solution set of $\dfrac{1}{2} < x$ and $x < 1$ is the interval $(1/2, 1)$ and the graph is

29. The solution set of $x > -4$ or $x > -1$ is the interval $(-4, \infty)$ and the graph is

31. Solving an equivalent statement, we get

$$x - 3 > 2 \quad \text{or} \quad x - 3 < -2$$
$$x > 5 \quad \text{or} \quad x < 1.$$

The solution set is

$$(-\infty, 1) \cup (5, \infty)$$

and the graph is

33. Since an absolute value is nonnegative, $2x - 7 = 0$. The solution set is $\{7/2\}$ and the graph is

35. Since absolute values are nonnegative, the solution set is $(-\infty, \infty)$ and

the graph is

37. Circle with radius 5 and center at the origin.

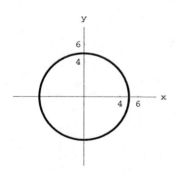

39. Equivalently, by using the method of completing the square, the circle is given by $(x + 2)^2 + y^2 = 4$. It has radius 2 and center $(-2, 0)$.

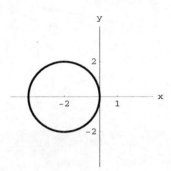

41. The line $y = -x + 25$ has intercepts $(0, 25)$, $(25, 0)$.

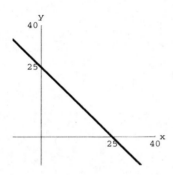

43. The line $y = 3x - 4$ has intercepts $(0, -4)$, $(4/3, 0)$.

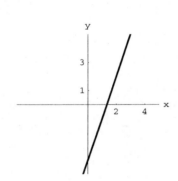

45. Vertical line $x = 5$ has intercept $(5, 0)$.

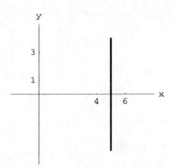

47. The distance is

$$\sqrt{(2 - (-3))^2 + (4 - 1)^2} = \sqrt{25 + 9} = \sqrt{34}.$$

49. Simplify $(x - (-3))^2 + (y - 5)^2 = \left(\sqrt{3}\right)^2$.
The standard equation is $(x+3)^2 + (y-5)^2 = 3$.

51. Substitute $y = 0$ in $3x - 4y = 12$. Then $3x = 12$ or $x = 4$. The x-intercept is $(4, 0)$. Substitute $x = 0$ in $3x - 4y = 12$ to get $-4y = 12$ or $y = -3$. The y-intercept is $(0, -3)$.

53. $\dfrac{2 - (-6)}{-1 - 3} = \dfrac{8}{-4} = -2$

55. Note, $m = \dfrac{-1 - 3}{5 - (-2)} = -\dfrac{4}{7}$. Solving for y in

$y - 3 = -\dfrac{4}{7}(x + 2)$, we obtain $y = -\dfrac{4}{7}x + \dfrac{13}{7}$.

57. Note, the slope of $3x + y = -5$ is -3.
The standard form for the line through $(2, -4)$ with slope $\dfrac{1}{3}$ is derived below.

$$\begin{aligned}
y + 4 &= \frac{1}{3}(x - 2) \\
3y + 12 &= x - 2 \\
-x + 3y &= -14 \\
x - 3y &= 14
\end{aligned}$$

59. Function, domain and range are both $\{-2, 0, 1\}$

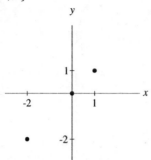

61. $y = 3 - x$ is a function, domain and range are both $(-\infty, \infty)$

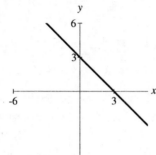

63. Not a function, domain is $\{2\}$, range is $(-\infty, \infty)$

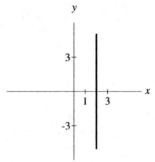

65. $x^2 + y^2 = 0.01$ is not a function, domain and range are both $[-0.1, 0.1]$

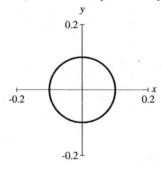

67. $x = y^2 + 1$ is not a function, domain is $[1, \infty)$, range is $(-\infty, \infty)$

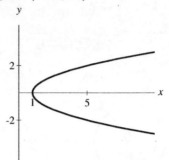

69. $y = \sqrt{x} - 3$ is a function, domain is $[0, \infty)$, range is $[-3, \infty)$

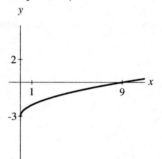

71. $9 + 3 = 12$

73. $24 - 7 = 17$

75. If $x^2 + 3 = 19$, then $x^2 = 16$ or $x = \pm 4$.

77. $g(12) = 17$

79. $7 + (-3) = 4$

81. $(4)(-9) = -36$

83. $f(-3) = 12$

85.

$$
\begin{aligned}
f(g(x)) &= f(2x - 7) \\
&= (2x - 7)^2 + 3 \\
&= 4x^2 - 28x + 52
\end{aligned}
$$

87. $(x^2 + 3)^2 + 3 = x^4 + 6x^2 + 12$

89. $(a + 1)^2 + 3 = a^2 + 2a + 4$

91.

$$\frac{f(3+h)-f(3)}{h} = \frac{(9+6h+h^2)+3-12}{h}$$
$$= \frac{6h+h^2}{h}$$
$$= 6+h$$

93.

$$\frac{f(x+h)-f(x)}{h} =$$
$$\frac{(x^2+2xh+h^2)+3-x^2-3}{h} =$$
$$\frac{2xh+h^2}{h} =$$
$$2x+h =$$

95. $g\left(\dfrac{x+7}{2}\right) = (x+7)-7 = x$

97. $g^{-1}(x) = \dfrac{x+7}{2}$

99. $f(x) = \sqrt{x}, g(x) = 2\sqrt{x+3}$; left by 3, stretch by 2

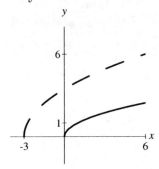

101. $f(x) = |x|, g(x) = -2|x+2|+4$; left by 2, stretch by 2, reflect about x-axis, up by 4

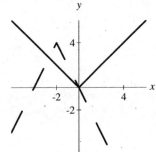

103. $f(x) = x^2, g(x) = \dfrac{1}{2}(x-2)^2 + 1$; right by 2, stretch by $\dfrac{1}{2}$, up by 1

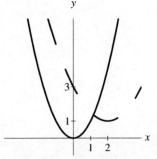

105. Translate the graph of f to the right by 2-units, stretch by a factor of 2, shift up by 1-unit.

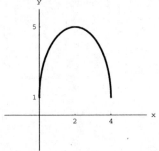

107. Translate the graph of f to the left by 1-unit, reflect about the x-axis, shift down by 3-units.

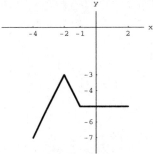

109. Translate the graph of f to the left by 2-units, stretch by a factor of 2, reflect about the x-axis.

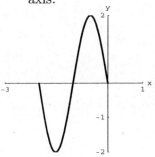

111. Stretch the graph of f by a factor of 2, reflect about the x-axis, shift up by 3-units.

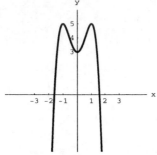

113. $F = f \circ g$

115. $H = f \circ h \circ g \circ j$

117. $N = h \circ f \circ j$

119. $R = g \circ h \circ j$

121.

$$\frac{f(x+h) - f(x)}{h} = \frac{-5(x+h) + 9 + 5x - 9}{h}$$

$$= \frac{-5h}{h}$$

$$= -5$$

123.

$$\frac{f(x+h) - f(x)}{h} =$$

$$= \frac{\dfrac{1}{2x+2h} - \dfrac{1}{2x}}{h} \cdot \frac{(2x+2h)(2x)}{(2x+2h)(2x)}$$

$$= \frac{(2x) - (2x+2h)}{h(2x+2h)(2x)}$$

$$= \frac{-2}{(2x+2h)(2x)}$$

$$= \frac{-1}{(x+h)(2x)}$$

125. Domain is $[-10, 10]$, range is $[0, 10]$, increasing on $(-10, 0)$, decreasing on $(0, 10)$

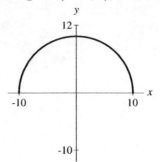

127. Domain and range are both $(-\infty, \infty)$, increasing on $(-\infty, \infty)$

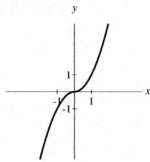

129. Domain is $(-\infty, \infty)$, range is $[-2, \infty)$, increasing on $(-2, 0)$ and $(2, \infty)$, decreasing on $(-\infty, -2)$ and $(0, 2)$

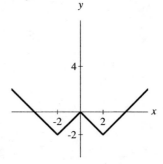

131. $y = |x| - 3$, domain is $(-\infty, \infty)$, range is $[-3, \infty)$

133. $y = -2|x| + 4$, domain is $(-\infty, \infty)$, range is $(-\infty, 4]$

135. $y = |x + 2| + 1$, domain is $(-\infty, \infty)$, range is $[1, \infty)$

137. Symmetry: y-axis

139. Symmetric about the origin

141. Neither symmetry

143. Symmetric about the y-axis

145. Inverse functions,
$f(x) = \sqrt{x+3}, g(x) = x^2 - 3$ for $x \geq 0$

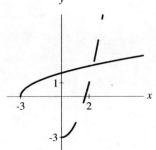

147. Inverse functions,

$$f(x) = 2x - 4, g(x) = \frac{1}{2}x + 2$$

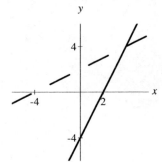

149. Not invertible

151. Inverse is $f^{-1}(x) = \dfrac{x + 21}{3}$ with domain and range both $(-\infty, \infty)$

153. Not invertible

155. Inverse is $f^{-1}(x) = x^2 + 9$ for $x \geq 0$ with domain $[0, \infty)$ and range $[9, \infty)$

157. Inverse is $f^{-1}(x) = \dfrac{5x + 7}{1 - x}$ with domain $(-\infty, 1) \cup (1, \infty)$, and range $(-\infty, -5) \cup (-5, \infty)$

159. Inverse is $f^{-1}(x) = -\sqrt{x - 1}$ with domain $[1, \infty)$ and range $(-\infty, 0]$

161. Let x be the number of roses. The cost function is $C(x) = 1.20x + 40$, the revenue function is $R(x) = 2x$, and the profit function is $P(x) = R(x) - C(x)$ or $P(x) = 0.80x - 40$.

Since $P(50) = 0$, to make a profit she must sell at least 51 roses.

163. Since $h(0) = 64$ and $h(2) = 0$, the range of $h = -16t^2 + 64$ is the interval $[0, 64]$. Then the domain of the inverse function is $[0, 64]$. Solving for t, we obtain

$$
\begin{aligned}
16t^2 &= 64 - h \\
t^2 &= \frac{64 - h}{16} \\
t &= \frac{\sqrt{64 - h}}{4}.
\end{aligned}
$$

The inverse function is $t = \dfrac{\sqrt{64 - h}}{4}$.

165. Since $A = \pi \left(\dfrac{d}{2}\right)^2$, $d = 2\sqrt{\dfrac{A}{\pi}}$.

167. The average rate of change is
$$\frac{8 - 6}{4} = 0.5 \text{ inch/lb.}$$

For Thought

1. False, the range of $y = x^2$ is $[0, \infty)$.

2. False, the vertex is the point $(3, -1)$.

3. True **4.** True **5.** True, since $\dfrac{-b}{2a} = \dfrac{6}{2 \cdot 3} = 1$.

6. True, the x-intercept of $y = (3x + 2)^2$ is the vertex $(-2/3, 0)$ and the y-intercept is $(0, 4)$.

7. True

8. True, since $(x - \sqrt{3})^2$ is always nonnegative.

9. True, since if x and $\dfrac{p - 2x}{2}$ are the length and the width, respectively, of a rectangle with perimeter p, then the area is $y = x \cdot \dfrac{p - 2x}{2}$. This is a parabola opening down with vertex $\left(\dfrac{p}{4}, \dfrac{p^2}{16}\right)$. Thus, the maximum area is $\dfrac{p^2}{16}$.

10. False

2.1 Exercises

1. Completing the square, we get

$$
\begin{aligned}
y &= \left(x^2 + 4x + \left(\frac{4}{2}\right)^2\right) - \left(\frac{4}{2}\right)^2 \\
y &= \left(x^2 + 4x + 4\right) - 4 \\
y &= (x + 2)^2 - 4.
\end{aligned}
$$

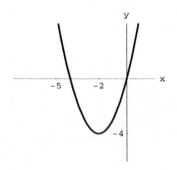

3. $y = \left(x^2 - 3x + \dfrac{9}{4}\right) - \dfrac{9}{4} = \left(x - \dfrac{3}{2}\right)^2 - \dfrac{9}{4}$

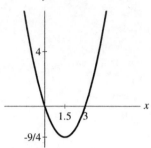

5. Completing the square, we get

$$
\begin{aligned}
y &= 2\left(x^2 - 6x + \left(\frac{6}{2}\right)^2\right) - 2\left(\frac{6}{2}\right)^2 + 22 \\
y &= 2(x - 3)^2 + 4.
\end{aligned}
$$

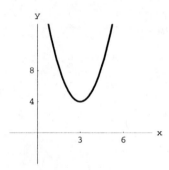

7. Completing the square, we find

$$
\begin{aligned}
y &= -3\left(x^2 - 2x + \left(\frac{2}{2}\right)^2\right) + 3\left(\frac{2}{2}\right)^2 - 3 \\
y &= -3(x - 1)^2.
\end{aligned}
$$

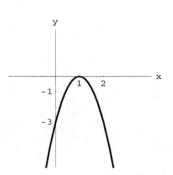

9. $y = \left(x^2 + 3x + \dfrac{9}{4}\right) - \dfrac{9}{4} + \dfrac{5}{2} = \left(x + \dfrac{3}{2}\right)^2 + \dfrac{1}{4}$

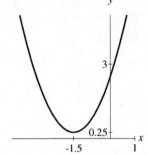

11. $y = -2\left(x^2 - \dfrac{3}{2}x + \dfrac{9}{16}\right) + \dfrac{9}{8} - 1 =$

$-2\left(x - \dfrac{3}{4}\right)^2 + \dfrac{1}{8}$

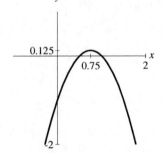

13. Since $\dfrac{-b}{2a} = \dfrac{12}{6} = 2$ and $f(2) = 12 - 24 + 1 = -11$, the vertex is $(2, -11)$.

15. Vertex: $(4, 1)$

17. Since $\dfrac{-b}{2a} = \dfrac{1/3}{-1} = -\dfrac{1}{3}$ and $f(-1/3) =$

$-\dfrac{1}{18} + \dfrac{1}{9} = \dfrac{1}{18}$, the vertex is $\left(-\dfrac{1}{3}, \dfrac{1}{18}\right)$.

19. Up, vertex $(1, -4)$, axis of symmetry $x = 1$, range $[-4, \infty)$, minimum value -4, decreasing on $(-\infty, 1)$, inreasing on $(1, \infty)$.

21. Since it opens down with vertex $(0, 3)$, the range is $(-\infty, 3]$, maximum value is 3, decreasing on $(0, \infty)$, and increasing on $(-\infty, 0)$.

23. Since it opens up with vertex $(1, -1)$, the range is $[-1, \infty)$, minimum value is -1, decreasing on $(-\infty, 1)$, and increasing on $(1, \infty)$.

25. Since it opens up with vertex $(-4, -18)$, range is $[-18, \infty)$, minimum value is -18, decreasing on $(-\infty, -4)$, and increasing on $(-4, \infty)$.

27. Since it opens up with vertex is $(3, 4)$, the range is $[4, \infty)$, minimum value is 4, decreasing on $(-\infty, 3)$, and increasing on $(3, \infty)$.

29. Since it opens down with vertex $(3/2, 27/2)$, the range is $(-\infty, 27/2]$, maximum value is $27/2$, decreasing on $(3/2, \infty)$, and increasing on $(-\infty, 3/2)$.

31. Since it opens down with vertex is $(1/2, 9)$, the range is $(-\infty, 9]$, maximum value is 9, decreasing on $(1/2, \infty)$, and increasing on $(-\infty, 1/2)$.

33. Vertex $(0, -3)$, axis $x = 0$, y-intercept $(0, -3)$, x-intercepts $(\pm\sqrt{3}, 0)$, opening up

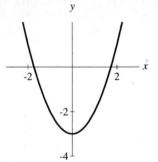

35. Vertex $(1/2, -1/4)$, axis $x = 1/2$, y-intercept $(0, 0)$, x-intercepts $(0, 0), (1, 0)$, opening up

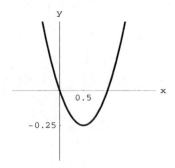

37. Vertex $(-3, 0)$, axis $x = -3$, y-intercept $(0, 9)$, x-intercept $(-3, 0)$, opening up

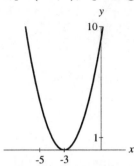

39. Vertex $(3, -4)$, axis $x = 3$, y-intercept $(0, 5)$, x-intercepts $(1, 0)$, $(5, 0)$, opening up

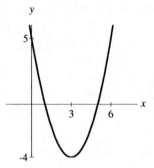

41. Vertex $(2, 12)$, axis $x = 2$, y-intercept $(0, 0)$, x-intercepts $(0, 0)$, $(4, 0)$, opening down

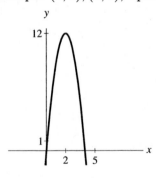

43. Vertex $(1, 3)$, axis $x = 1$, y-intercept $(0, 1)$, x-intercepts $\left(1 \pm \dfrac{\sqrt{6}}{2}, 0\right)$, opening down

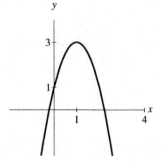

45. The sign graph of $(2x - 3)(x + 1) < 0$ is

```
- - - - - - - -  0 + + + +
- - - - 0 + + + + + + + +
←——————————————————————→
       -1          3/2
```

The solution set is the interval $(-1, 3/2)$

and the graph is
```
        -1    3/2
←————————(═════)————→
```

47. The sign graph of $(x + 3)(x - 5) > 0$ is

```
- - - - - - - -  0 + + + +
- - - - 0 + + + + + + + +
←——————————————————————→
       -3          5
```

The solution set is $(-\infty, -3) \cup (5, \infty)$

and the graph is
```
     -3   5
←════)———(════→
```

49. The sign graph of $(w + 2)(w - 6) \geq 0$ is

```
- - - - - - - -  0 + + + +
- - - - 0 + + + + + + + +
←——————————————————————→
       -2          6
```

The solution set is $(-\infty, -2] \cup [6, \infty)$

and the graph is
```
     -2  6
←════]——[════→
```

51. The sign graph of $(t - 4)(t + 4) \leq 0$ is

```
- - - - - - - - -  0 + + + +
- - - - 0 + + + + + + + +
←——————————————————————→
       -4          4
```

The solution set is $[-4, 4]$ and the

graph is
```
        -4    4
←————————[═══]————→
```

53. The sign graph of $(a + 3)^2 \leq 0$ is

```
+ + + + 0 + + + + + + + + +
←——————————————————————→
       -3
```

The solution set is $\{-3\}$ and the

graph is
```
          -3
←————————•————————→
```

55. The sign graph of $(2z - 3)^2 > 0$ is

```
+ + + + 0 + + + + + + + + +
←——————————————————————→
       3/2
```

The solution set is $(-\infty, 3/2) \cup (3/2, \infty)$

and the graph is

$$\xleftarrow{\hspace{2cm}}\overset{3/2}{)(}\xrightarrow{\hspace{2cm}}$$

57. The roots of $x^2 - 4x + 2 = 0$ are $x_1 = 2 - \sqrt{2}$ and $x_2 = 2 + \sqrt{2}$.

If $x = -5$, then $(-5)^2 - 4(-5) + 2 > 0$.
If $x = 2$, then $(2)^2 - 4(2) + 2 < 0$.
If $x = 5$, then $(5)^2 - 4(5) + 2 > 0$.

$$\begin{array}{ccccc} + & 0 & - & 0 & + \\ \hline -5 & x_1 & 0 & x_2 & 5 \end{array}$$

The solution set of $x^2 - 4x + 2 < 0$

is $(2 - \sqrt{2}, 2 + \sqrt{2})$ and its graph follows.

$$\xleftarrow{\hspace{1cm}}\overset{2-\sqrt{2} \quad 2+\sqrt{2}}{(\underline{\qquad})}\xrightarrow{\hspace{1cm}}$$

59. The roots of $x^2 - 10 = 0$ are $x_1 = -\sqrt{10}$ and $x_2 = \sqrt{10}$.

If $x = -4$, then $(-4)^2 - 10 > 0$.
If $x = 0$, then $(0)^2 - 10 < 0$.
If $x = 4$, then $(4)^2 - 10 > 0$.

$$\begin{array}{ccccc} + & 0 & - & 0 & + \\ \hline -4 & x_1 & 0 & x_2 & 4 \end{array}$$

The solution set of $x^2 - 9 \geq 1$

is $(-\infty, -\sqrt{10}] \cup [\sqrt{10}, \infty)$ and its

graph is $\xleftarrow{\hspace{1cm}}\overset{-\sqrt{10} \quad \sqrt{10}}{\underline{\quad}] \quad [\underline{\quad}}\xrightarrow{\hspace{1cm}}$

61. The roots of $y^2 - 10y + 18 = 0$ are $y_1 = 5 - \sqrt{7}$ and $y_2 = 5 + \sqrt{7}$.

If $y = 2$, then $(2)^2 - 10(2) + 18 > 0$.
If $y = 5$, then $(5)^2 - 10(5) + 18 < 0$.
If $y = 8$, then $(8)^2 - 10(8) + 18 > 0$.

$$\begin{array}{ccccc} + & 0 & - & 0 & + \\ \hline 2 & y_1 & 5 & y_2 & 8 \end{array}$$

The solution set of $y^2 - 10y + 18 > 0$

is $(-\infty, 5 - \sqrt{7}) \cup (5 + \sqrt{7}, \infty)$ and its

graph is $\xleftarrow{\hspace{1cm}}\overset{5-\sqrt{7} \quad 5+\sqrt{7}}{) \quad (}\xrightarrow{\hspace{1cm}}$

63. Note, $p^2 + 9 = 0$ has no real roots.
If $p = 0$, then $(0)^2 + 9 > 0$. The signs of $p^2 + 9$ are shown below.

$$\begin{array}{c} + \\ \hline 0 \end{array}$$

The solution set of $p^2 + 9 > 0$ is $(-\infty, \infty)$

and its graph follows. $\xleftarrow{\hspace{2cm}}\xrightarrow{\hspace{2cm}}$

65. Note, $a^2 - 8a + 20 = 0$ has no real roots.
If $a = 0$, then $(0)^2 - 8(0) + 20 > 0$. The signs of $a^2 - 8a + 20$ are shown below.

$$\begin{array}{c} + \\ \hline 0 \end{array}$$

The solution set of $a^2 - 8a + 20 \leq 0$ is $\emptyset$.

67. Note, $2w^2 - 5w + 6 = 0$ has no real roots.
If $w = 0$, then $2(0)^2 - 5(0) + 6 > 0$. The signs of $2w^2 - 5w + 6$ are shown below.

$$\begin{array}{c} + \\ \hline 0 \end{array}$$

The solution set of $2w^2 - 5w + 6 > 0$ is $(-\infty, \infty)$

and its graph follows. $\xleftarrow{\hspace{2cm}}\xrightarrow{\hspace{2cm}}$

69. $(-\infty, -1] \cup [3, \infty)$

71. $(-3, 1)$

73. $[-3, 1]$

75. a) Since $x^2 - 3x - 10 = (x - 5)(x + 2) = 0$, the solution set is $\{-2, 5\}$.

b) Since $x^2 - 3x - 10 = -10$, we get $x^2 - 3x = 0$ or $x(x - 3) = 0$. The solution set is $\{0, 3\}$.

c) If $x = -3$, then $(-3)^2 - 3(-3) - 10 > 0$.
If $x = 0$, then $(0)^2 - 3(0) - 10 < 0$.
If $x = 6$, then $(6)^2 - 3(6) - 10 > 0$.
The signs of $x^2 - 3x - 10$ are shown below.

$$\begin{array}{ccccc} + & 0 & - & 0 & + \\ \hline -3 \quad -2 & 0 & & 5 & 6 \end{array}$$

The solution set of $x^2 - 3x - 10 > 0$.
is $(-\infty, -2) \cup (5, \infty)$

d) Using the sign graph of $x^2 - 3x - 10$ given in part c), the solution set of $x^2 - 3x - 10 \leq 0$ is $[-2, 5]$.

e) By using the method of completing the square, one obtains

$$x^2 - 3x - 10 \;=\; \left(x - \frac{3}{2}\right)^2 - 10 - \frac{9}{4}$$

$$x^2 - 3x - 10 \;=\; \left(x - \frac{3}{2}\right)^2 - \frac{49}{4}.$$

The graph of f is obtained from the graph of $y = x^2$ by shifting to the right by $\dfrac{3}{2}$ unit, and down by $\dfrac{49}{4}$.

f) Domain is $(-\infty, \infty)$, range is $\left[-\dfrac{49}{4}, \infty\right)$, minimum y-value is $-\dfrac{49}{4}$

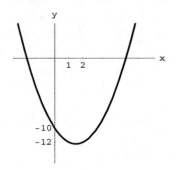

g) The solution to $f(x) > 0$ may be obtained by considering the part of the parabola that is above the x-axis, i.e., when x lies in $(-\infty, -2) \cup (5, \infty)$.

While, the solution to $f(x) \leq 0$ may be obtained by considering the part of the parabola on or below the x-axis. i.e., when x is in $[-2, 5]$.

h) x-intercepts are $(5, 0)$ and $(-2, 0)$, the y-intercept is $(0, -10)$, axis of symmetry $x = \dfrac{3}{2}$, vertex $\left(\dfrac{3}{2}, -\dfrac{49}{4}\right)$, opens up, increasing on $\left(\dfrac{3}{2}, \infty\right)$, decreasing on $\left(-\infty, \dfrac{3}{2}\right)$

77. Since $\dfrac{-b}{2a} = \dfrac{-128}{-2(-16)} = 4$, the maximum height is $h(4) = 261$ ft.

79. a) Finding the vertex involves the number

$$\frac{-b}{2a} = \frac{-160}{-32} = 5.$$

Thus, the maximum height is

$$h(5) = -16(5)^2 + 160(5) + 8 = 408 \text{ ft.}$$

b) When the arrow reaches the ground, one has $h(t) = 0$.

$$-16t^2 + 160t + 8 \;= 0$$
$$t^2 - 10t \;=\; \frac{1}{2}$$
$$(t - 5)^2 \;=\; 25 + \frac{1}{2}$$
$$t \;=\; 5 \pm \sqrt{\frac{51}{2}}$$

Since $t \geq 0$, the arrow reaches the ground in

$$5 + \frac{\sqrt{102}}{2} = \frac{10 + \sqrt{102}}{2} \approx 10.05 \text{ sec.}$$

81. Let x and y be the length and width, respectively. Since $2x + 2y = 200$, we find $y = 100 - x$. The area as a function of x is

$$f(x) = x(100 - x) = 100x - x^2.$$

The graph of f is a parabola and its vertex is $(50, 2500)$. Thus, the maximum area is 2500 yd^2. Using $x = 50$ from the vertex, we get $y = 100 - x = 100 - 50 = 50$. The dimensions are 50 yd by 50 yd.

83. Let the length of the sides be x, x, x, y, y.

Then $3x + 2y = 120$ and $y = \dfrac{120 - 3x}{2}$.

The area of rectangular enclosure is

$$A(x) = xy = x\left(\frac{120 - 3x}{2}\right) = \frac{1}{2}(120x - 3x^2).$$

This is a parabola opening down. Since

$$-\frac{b}{2a} = 20 \text{ and } y = \frac{120 - 3(20)}{2} = 30, \text{ the}$$

optimal dimensions are 20 ft by 30 ft.

85. Let the length of the sides be x, x, and $30-2x$. The area of rectangular enclosure is

$$A(x) = x(30 - 2x) = 30x - 2x^2.$$

We have a parabola opening down. Since $-\dfrac{b}{2a} = 7.5$ and $y = 30 - 2(7.5) = 15$, the optimal dimensions are 15 ft by 7.5 ft.

87. Let n and p be the number of persons and the price of a tour per person, respectively.

a) The function expressing p as a function of n is $p = 50 - n$.

b) The revenue is $R = (50 - n)n$ or

$$R = 50n - n^2.$$

c) Since the graph of R is a parabola opening down with vertex $(25, 625)$, we find that 25 persons will give her the maximum revenue of \$625.

For Thought

1. True, since $i \cdot (-i) = 1$.

2. True, since $\overline{0 + i} = 0 - i = -i$.

3. False, the set of real numbers is a subset of the complex numbers.

4. True, $(\sqrt{3} - i\sqrt{2})(\sqrt{3} + i\sqrt{2}) = 3 + 2 = 5$.

5. False, since $(2 + 5i)(2 + 5i) = 4 + 20i + 25i^2 = 4 + 20i - 25 = -21 + 20i$.

6. False, $5 - \sqrt{-9} = 5 - 3i$.

7. True, since $(3i)^2 + 9 = (-9) + 9 = 0$.

8. True, since $(-3i)^2 + 9 = (-9) + 9 = 0$.

9. True, since $i^4 = i^2 \cdot i^2 = (-1)(-1) = 1$.

10. False, $i^{18} = (i^4)^4 i^2 = (1)^4(-1) = -1$.

2.2 Exercises

1. $0 + 6i$, imaginary

3. $\dfrac{1}{3} + \dfrac{1}{3}i$, imaginary

5. $\sqrt{7} + 0i$, real

7. $\dfrac{\pi}{2} + 0i$, real

9. $7 + 2i$

11. $1 - i - 3 - 2i = -2 - 3i$

13. $-18i + 12i^2 = -12 - 18i$

15. $8 + 12i - 12i - 18i^2 = 26$

17. $(5 - 2i)(5 + 2i) = 25 - 4i^2 = 25 - 4(-1) = 29$

19. $(\sqrt{3} - i)(\sqrt{3} + i) = 3 - i^2 = 3 - (-1) = 4$

21. $9 + 24i + 16i^2 = -7 + 24i$

23. $5 - 4i\sqrt{5} + 4i^2 = 1 - 4i\sqrt{5}$

25. $(i^4)^4 \cdot i = (1)^4 \cdot i = i$

27. $(i^4)^{24} i^2 = 1^{24}(-1) = -1$

29. $(i^4)^{-1} = 1^{-1} = 1$

31. Since $i^4 = 1$, we get $i^{-1} = i^{-1}i^4 = i^3 = -i$.

33. $(3 - 9i)(3 + 9i) = 9 - 81i^2 = 90$

35. $\left(\dfrac{1}{2} + 2i\right)\left(\dfrac{1}{2} - 2i\right) = \dfrac{1}{4} - 4i^2 = \dfrac{1}{4} + 4 = \dfrac{17}{4}$

37. $i(-i) = -i^2 = 1$

39. $(3 - i\sqrt{3})(3 + i\sqrt{3}) = 9 - 3i^2 = 9 - 3(-1) = 12$

41. $\dfrac{1}{2 - i} \cdot \dfrac{2 + i}{2 + i} = \dfrac{2 + i}{5} = \dfrac{2}{5} + \dfrac{1}{5}i$

43. $\dfrac{-3i}{1 - i} \cdot \dfrac{1 + i}{1 + i} = \dfrac{-3i + 3}{2} = \dfrac{3}{2} - \dfrac{3}{2}i$

45. $-1 + 3i$

47.
$$\dfrac{-3 + 3i}{i} \cdot \dfrac{-i}{-i} = \dfrac{3i - 3i^2}{1} = 3i - 3(-1) = 3 + 3i$$

49.
$$\dfrac{1 - i}{3 + 2i} \cdot \dfrac{3 - 2i}{3 - 2i} = \dfrac{3 - 5i - 2}{13} = \dfrac{1}{13} - \dfrac{5}{13}i$$

51. $2i - 3i = -i$

53. $-4 + 2i$

55. $\left(i\sqrt{6}\right)^2 = -6$

57. $(i\sqrt{2})(i\sqrt{50}) = i^2\sqrt{2}\cdot 5\sqrt{2} = (-1)(2)(5) = -10$

59.
$$\frac{-2}{2} + \frac{i\sqrt{20}}{2} = -1 + i\frac{2\sqrt{5}}{2} = -1 + \sqrt{5}i$$

61. $-3 + \sqrt{9 - 20} = -3 + i\sqrt{11}$

63. $2i\sqrt{2}\left(i\sqrt{2} + 2\sqrt{2}\right) = 4i^2 + 8i = -4 + 8i$

65. $\dfrac{-2 + \sqrt{-16}}{2} = \dfrac{-2 + 4i}{2} = -1 + 2i$

67. $\dfrac{-4 + \sqrt{16 - 24}}{4} = \dfrac{-4 + 2\sqrt{2}i}{4} = \dfrac{-2 + i\sqrt{2}}{2}$

69. $\dfrac{-6 - \sqrt{-32}}{2} = \dfrac{-6 - 4i\sqrt{2}}{2} = -3 - 2i\sqrt{2}$

71. $\dfrac{-6 - \sqrt{36 + 48}}{-4} = \dfrac{-6 - 2\sqrt{21}}{-4} = \dfrac{3 + \sqrt{21}}{2}$

73. Since $x^2 = -1$, the solution set is $\{\pm i\}$.

75. Since $x^2 = -8$, the solution set is $\left\{\pm 2i\sqrt{2}\right\}$.

77. Since $x^2 = -\dfrac{1}{2}$, the solution set is $\left\{\pm i\dfrac{\sqrt{2}}{2}\right\}$.

79. Completing the square, we find

$$(x-1)^2 + 1 = 0.$$

Then the solution set is $\{1 \pm i\}$.

81. Completing the square, we find

$$(x-2)^2 + 9 = 0.$$

Then the solution set is $\{2 \pm 3i\}$.

83. Note, $a = 1, b = -2$, and $c = 4$. Then

$$x = \frac{2 \pm \sqrt{4 - 16}}{2} = \frac{2 \pm 2i\sqrt{3}}{2}.$$

The solution set is $\left\{1 \pm i\sqrt{3}\right\}$.

85. Since $2x^2 - 2x + 5 = 0$, we find $a = 2, b = -2$, and $c = 5$. Then

$$x = \frac{2 \pm \sqrt{4 - 40}}{4} = \frac{2 \pm 6i}{4}.$$

The solution set is $\left\{\dfrac{1}{2} \pm \dfrac{3}{2}i\right\}$.

87. Since $a = 4, b = -8, c = 7$ and

$$x = \frac{8 \pm \sqrt{64 - 112}}{8} = \frac{8 \pm \sqrt{-48}}{8} =$$

$$\frac{8 \pm 4i\sqrt{3}}{8}, \text{ the solution set is } \left\{1 \pm \frac{\sqrt{3}}{2}i\right\}.$$

For Thought

1. False   **2.**   True, by the Remainder Theorem.

3. True, by the Factor Theorem.

4. True, since $2^5 - 1 = 31$.

5. False, since $P(x) = 1$ has no zero.

6. False, rather $c^3 - c^2 + 4c - 5 = b$.

7. False, since $P(4) = -15$.   **8.**   True

9. True, since 1 is a root.

10. False, since -3 is not a root.

2.3 Exercises

1. Quotient $x - 3$, remainder 1

$$
\begin{array}{r}
x - 3 \\
x - 2 \overline{\smash{\big)}\ x^2 - 5x + 7} \\
\underline{x^2 - 2x} \\
-3x + 7 \\
\underline{-3x + 6} \\
1
\end{array}
$$

3. Quotient $-2x^2 + 6x - 14$, remainder 33

$$
\begin{array}{r}
-2x^2 + 6x - 14 \\
x + 3 \overline{\smash{\big)}\ -2x^3 + 0x^2 + 4x - 9} \\
\underline{-2x^3 - 6x^2} \\
6x^2 + 4x \\
\underline{6x^2 + 18x} \\
-14x - 9 \\
\underline{-14x - 42} \\
33
\end{array}
$$

5. Quotient $s^2 + 2$, remainder 16

$$
\begin{array}{r}
s^2 + 2 \\
s^2 - 5 \overline{)s^4 - 3s^2 + 6} \\
\underline{s^4 - 5s^2} \\
2s^2 + 6 \\
\underline{2s^2 - 10} \\
16
\end{array}
$$

7. Quotient $x + 6$, remainder 13

$$
\begin{array}{c|ccc}
2 & 1 & 4 & 1 \\
& & 2 & 12 \\
\hline
& 1 & 6 & 13
\end{array}
$$

9. Quotient $-x^2 + 4x - 16$, remainder 57

$$
\begin{array}{c|cccc}
-3 & -1 & 1 & -4 & 9 \\
& & 3 & -12 & 48 \\
\hline
& -1 & 4 & -16 & 57
\end{array}
$$

11. Quotient $4x^2 + 2x - 4$, remainder 0

$$
\begin{array}{c|cccc}
1/2 & 4 & 0 & -5 & 2 \\
& & 2 & 1 & -2 \\
\hline
& 4 & 2 & -4 & 0
\end{array}
$$

13. Quotient $2a^2 - 4a + 6$, remainder 0

$$
\begin{array}{c|cccc}
-1/2 & 2 & -3 & 4 & 3 \\
& & -1 & 2 & -3 \\
\hline
& 2 & -4 & 6 & 0
\end{array}
$$

15. Quotient $x^3 + x^2 + x + 1$, remainder -2

$$
\begin{array}{c|ccccc}
1 & 1 & 0 & 0 & 0 & -3 \\
& & 1 & 1 & 1 & 1 \\
\hline
& 1 & 1 & 1 & 1 & -2
\end{array}
$$

17. Quotient $x^4 + 2x^3 - 2x^2 - 4x - 4$, remainder -13

$$
\begin{array}{c|cccccc}
2 & 1 & 0 & -6 & 0 & 4 & -5 \\
& & 2 & 4 & -4 & -8 & -8 \\
\hline
& 1 & 2 & -2 & -4 & -4 & -13
\end{array}
$$

19. $f(1) = 0$

$$
\begin{array}{c|cccccc}
1 & 1 & 0 & 0 & 0 & 0 & -1 \\
& & 1 & 1 & 1 & 1 & 1 \\
\hline
& 1 & 1 & 1 & 1 & 1 & 0
\end{array}
$$

21. $f(-2) = -33$

$$
\begin{array}{c|cccccc}
-2 & 1 & 0 & 0 & 0 & 0 & -1 \\
& & -2 & 4 & -8 & 16 & -32 \\
\hline
& 1 & -2 & 4 & -8 & 16 & -33
\end{array}
$$

23. $g(1) = 5$

$$
\begin{array}{c|cccc}
1 & 1 & -4 & 0 & 8 \\
& & 1 & -3 & -3 \\
\hline
& 1 & -3 & -3 & 5
\end{array}
$$

25. $g(-1/2) = 55/8$

$$
\begin{array}{c|cccc}
-1/2 & 1 & -4 & 0 & 8 \\
& & -1/2 & 9/4 & -9/8 \\
\hline
& 1 & -9/2 & 9/4 & 55/8
\end{array}
$$

27. $h(-1) = 0$

$$
\begin{array}{c|ccccc}
-1 & 2 & 1 & -1 & 3 & 3 \\
& & -2 & 1 & 0 & -3 \\
\hline
& 2 & -1 & 0 & 3 & 0
\end{array}
$$

29. $h(1) = 8$

$$
\begin{array}{c|ccccc}
1 & 2 & 1 & -1 & 3 & 3 \\
& & 2 & 3 & 2 & 5 \\
\hline
& 2 & 3 & 2 & 5 & 8
\end{array}
$$

31. Yes, $(x+3)(x^2+x-2) = (x+3)(x+2)(x-1)$

$$
\begin{array}{c|cccc}
-3 & 1 & 4 & 1 & -6 \\
& & -3 & -3 & 6 \\
\hline
& 1 & 1 & -2 & 0
\end{array}
$$

33. Yes, $(x-4)(x^2+8x+15) = (x-4)(x+5)(x+3)$

4	1	4	-17	-60
		4	32	60
	1	8	15	0

35. Yes, since the remainder below is zero

3	2	-5	-4	3
		6	3	-3
	2	1	-1	0

37. No, since the remainder below is not zero

-2	1	2	3	1
		-2	0	-6
	1	0	3	-5

39. Yes, since the remainder below is zero

-1	1	2	4	6	3
		-1	-1	-3	-3
	1	1	3	3	0

41. No, since the remainder below is not zero

1/2	1	3	-5	7
		1/2	7/4	-13/8
	1	7/2	-13/4	43/8

43. $\pm\{1,2,3,4,6,8,12,24\}$

45. $\pm\{1,3,5,15\}$

47. $\pm\left\{1,3,5,15,\dfrac{1}{2},\dfrac{1}{4},\dfrac{1}{8},\dfrac{3}{2},\dfrac{3}{4},\right.$
$\left.\dfrac{3}{8},\dfrac{5}{2},\dfrac{5}{4},\dfrac{5}{8},\dfrac{15}{2},\dfrac{15}{4},\dfrac{15}{8}\right\}$

49. $\pm\left\{1,2,\dfrac{1}{2},\dfrac{1}{3},\dfrac{1}{6},\dfrac{1}{9},\dfrac{1}{18},\dfrac{2}{3},\dfrac{2}{9}\right\}$

51. Zeros are $2,3,4$ since

2	1	-9	26	-24
		2	-14	24
	1	-7	12	0

and $x^2 - 7x + 12 = (x-4)(x-3)$

53. Zeros are $-3, 2 \pm i$ since

-3	1	-1	-7	15
		-3	12	-15
	1	-4	5	0

$x^2 - 4x + 5 = (x-2)^2 + 1 = 0$ or $x - 2 = \pm i$

55. Zeros are $1/2, 3/2, 5/2$ since

1/2	8	-36	46	-15
		4	-16	15
	8	-32	30	0

and $8a^2 - 32a + 30 = 2(4a^2 - 16a + 15) =$
$2(2a - 3)(2a - 5)$

57. Zeros are $\dfrac{1}{2}, \dfrac{1 \pm i}{3}$ since

1/2	18	-21	10	-2
		9	-6	2
	18	-12	4	0

and the zeros of $18t^2 - 12t + 4$ are
(by the quadratic formula) $\dfrac{1 \pm i}{3}$

59. Zeros are $1, -2, \pm i$ since

1	1	1	-1	1	-2
		1	2	1	2
	1	2	1	2	0

-2	1	2	1	2
		-2	0	-2
	1	0	1	0

and the zeros of $w^2 + 1$ are $\pm i$

61. Zeros are $-1, \pm\sqrt{2}$ since

-1	1	2	-1	-4	-2
		-1	-1	2	2
	1	1	-2	-2	0

-1	1	1	-2	-2
		-1	0	2
	1	0	-2	0

and the zeros of $x^2 - 2$ are $\pm\sqrt{2}$

63. Zeros are $1/2, 1/3, 1/4$ since

1/2	24	-26	9	-1
		12	-7	1
	24	-14	2	0

and $2(12x^2 - 7x + 1) = 2(4x - 1)(3x - 1)$

65. Rational zero is $1/16$ since

1/16	16	-33	82	-5
		1	-2	5
	16	-32	80	0

and by the quadratic formula $16x^2 - 32x + 80$ has imaginary zeros $1 \pm 2i$.

67. Rational zeros are $7/3, -6/7$ since

7/3	21	-31	-21	-31	-42
		49	42	49	42
	21	18	21	18	0

-6/7	21	18	21	18
		-18	0	-18
	21	0	21	0

and $21x^2 + 21 = 0$ has imaginary zeros $\pm i$.

69. Dividing $x^3 + 6x^2 + 3x - 10$ by $x - 1$, we find

1	1	6	3	-10
		1	7	10
	1	7	10	0

Moreover, $x^2 + 7x + 10 = (x + 5)(x + 2)$.
Note, the zeros of $x^2 + 9$ are $\pm 3i$.
Thus, the zeros of $f(x)$ are $x = -5, -2, 1, \pm 3i$.

71. Dividing $x^3 - 9x^2 + 23x - 15$ by $x - 1$, we find

1	1	-9	23	-15
		1	-8	15
	1	-8	15	0

For the quotient, we find
$$x^2 - 8x + 15 = (x - 5)(x - 3).$$
Then the zeros of $x^3 - 9x^2 + 23x - 15$ are $x = 1, 3, 5$.

By using the method of completing the square, we find
$$x^2 - 4x + 1 = (x - 2)^2 - 3.$$
Then the zeros of $(x - 2)^2 - 3$ are $2 \pm \sqrt{3}$.
Thus, all the zeros are $x = 1, 3, 5, 2 \pm \sqrt{3}$.

73.
$$\frac{2x + 1}{x - 2} = 2 + \frac{5}{x - 2} \text{ since}$$

2	2	1
		4
	2	5

75.
$$\frac{a^2 - 3a + 5}{a - 3} = a + \frac{5}{a - 3} \text{ since}$$

3	1	-3	5
		3	0
	1	0	5

77.
$$\frac{c^2 - 3c - 4}{c^2 - 4} = 1 + \frac{-3c}{c^2 - 4} \text{ since}$$

$$c^2 - 4 \enclose{longdiv}{c^2 - 3c - 4}$$
$$\underline{c^2 + 0c - 4}$$
$$-3c$$

79.
$$\frac{4t - 5}{2t + 1} = 2 + \frac{-7}{2t + 1} \text{ since}$$

$$2t + 1 \enclose{longdiv}{4t - 5} \quad 2$$
$$\underline{4t + 2}$$
$$-7$$

81. a) Note, $\dfrac{P(t)}{t} = -t^3 + 12t^2 - 58t + 132.$

$$
\begin{array}{r|rrrr}
6 & -1 & 12 & -58 & 132 \\
  &   & -6 & 36 & -132 \\
\hline
  & -1 & 6 & -22 & 0
\end{array}
$$

The drug will be eliminated in $t = 6$ hr.

b) About 120 ppm

c) About 3 hours

d) Between 1 and 5 hours approximately, the concentration is above 80 ppm. Thus, the concentration is above 80 ppm for about 4 hours.

83. If w is the width, then

$$w(w+4)(w+9) = 630.$$

This can be re-written as

$$w^3 + 13w^2 + 36w - 630 = 0.$$

Using synthetic division, we find

$$
\begin{array}{r|rrrr}
5 & 1 & 13 & 36 & -630 \\
  &   & 5 & 90 & 630 \\
\hline
  & 1 & 18 & 126 & 0
\end{array}
$$

Since $w^2 + 18w + 126 = 0$ has non-real roots, the width of the HP box is $w = 5$ in. The dimensions are 5 in. by 9 in. by 14 in.

For Thought

1. False, since 1 has multiplicity 1.   **2.**   True

3. True   **4.**   False, it factors as $(x-5)^4(x+2)$.

5. False, rather $4+5i$ is also a solution.   **6.**   True

7. False, since they are solutions to a polynomial with real coefficients of degree at least 4.

8. False, 2 is not a solution.

9. True, since $-x^3 - 5x^2 - 6x - 1 = 0$ has no sign changes.

10. True

2.4 Exercises

1. Degree 2; 5 with multiplicity 2 since $(x-5)^2 = 0$

3. Degree 5; 0 with multiplicity 3, and ± 3 since $x^3(x-3)(x+3) = 0$

5. Degree 4; 0 and 1 each have multiplicity 2 since $x^2(x^2 - 2x + 1) = x^2(x-1)^2$

7. Degree 4; 3/2 and $-4/3$ each with multiplicity 2

9. Degree 3; the roots are $0, 2 \pm \sqrt{10}$ since $x(x^2 - 4x - 6) = x((x-2)^2 - 10) = 0$

11. $x^2 + 9$

13. $\left[(x-1) - \sqrt{2}\right]\left[(x-1) + \sqrt{2}\right] = (x-1)^2 - 2 = x^2 - 2x - 1$

15. $[(x-3) - 2i][(x-3) + 2i] = (x-3)^2 + 4 = x^2 - 6x + 13$

17. $(x-2)[(x-3) - 4i][(x-3) + 4i] = (x-2)[(x-3)^2 + 16] = x^3 - 8x^2 + 37x - 50$

19. $(x+3)(x-5) = 0$ or $x^2 - 2x - 15 = 0$

21. $(x+4i)(x-4i) = 0$ or $x^2 + 16 = 0$

23. $(x-(3-i))(x-(3+i)) = 0$ or $x^2 - 6x + 10 = 0$

25. $(x+2)(x-i)(x+i) = 0$ or $x^3 + 2x^2 + x + 2 = 0$

27. $x(x-i\sqrt{3})(x+i\sqrt{3}) = 0$ or $x^3 + 3x = 0$

29. $(x-3)[x-(1-i)][x-(1+i)] = 0$ or $x^3 - 5x^2 + 8x - 6 = 0$

31. $(x-1)(x-2)(x-3) = 0$ or $x^3 - 6x^2 + 11x - 6 = 0$

33. $(x-1)[x-(2-3i)][x-(2+3i)] = 0$ or $x^3 - 5x^2 + 17x - 13 = 0$

35. $(2x-1)(3x-1)(4x-1) = 0$ or $24x^3 - 26x^2 + 9x - 1 = 0$

37. $(x-i)(x+i)[x-(1+i)][x-(1-i)] = 0$ or $x^4 - 2x^3 + 3x^2 - 2x + 2 = 0$

39. $P(x) = x^3 + 5x^2 + 7x + 1$ has no sign change and $P(-x) = -x^3 + 5x^2 - 7x + 1$ has 3 sign changes. There are (a) 3 negative roots, or (b) 1 negative root & 2 imaginary roots.

41. $P(x) = -x^3 - x^2 + 7x + 6$ has 1 sign change and $P(-x) = x^3 - x^2 - 7x + 6$ has 2 sign changes. There are (a) 1 positive root and 2 negative roots, or (b) 1 positive root and 2 imaginary roots.

43. $P(y) = y^4 + 5y^2 + 7 = P(-y)$ has no sign change. There are 4 imaginary roots.

45. $P(t) = t^4 - 3t^3 + 2t^2 - 5t + 7$ has 4 sign changes and $P(-t) = t^4 + 3t^3 + 2t^2 + 5t + 7$ has no sign change. There are (a) 4 positive roots, or (b) 2 positive roots and 2 imaginary roots, or (c) 4 imaginary roots.

47. $P(x) = x^5 + x^3 + 5x$ and $P(-x) = -x^5 - x^3 - 5x$ have no sign changes; 4 imaginary roots and 0.

49. Roots are $1, 5, -2$ since

$$
\begin{array}{r|rrrr}
1 & 1 & -4 & -7 & 10 \\
  &   & 1 & -3 & -10 \\
\hline
  & 1 & -3 & -10 & 0
\end{array}
$$

and $x^2 - 3x - 10 = (x - 5)(x + 2)$

51. Roots are $-3, \dfrac{3 \pm \sqrt{13}}{2}$ since

$$
\begin{array}{r|rrrr}
-3 & 1 & 0 & -10 & -3 \\
   &   & -3 & 9 & 3 \\
\hline
   & 1 & -3 & -1 & 0
\end{array}
$$

and by the quadratic formula the roots of $x^2 - 3x - 1 = 0$ are $\dfrac{3 \pm \sqrt{13}}{2}$.

53. Roots are $2, -4, \pm i$ since

$$
\begin{array}{r|rrrrr}
2 & 1 & 2 & -7 & 2 & -8 \\
  &   & 2 & 8 & 2 & 8 \\
\hline
  & 1 & 4 & 1 & 4 & 0
\end{array}
$$

$$
\begin{array}{r|rrrr}
-4 & 1 & 4 & 1 & 4 \\
   &   & -4 & 0 & -4 \\
\hline
   & 1 & 0 & 1 & 0
\end{array}
$$

and the root of $x^2 + 1 = 0$ are $\pm i$.

55. Roots are $1/3, 1/2, -5$ since

$$
\begin{array}{r|rrrr}
1/3 & 6 & 25 & -24 & 5 \\
    &   & 2 & 9   & -5 \\
\hline
    & 6 & 27 & -15 & 0
\end{array}
$$

and $6x^2 + 27x - 15 = 3(2x - 1)(x + 5)$.

57. $1, -2$ each have multiplicity 2 since

$$
\begin{array}{r|rrrrr}
1 & 1 & 2 & -3 & -4 & 4 \\
  &   & 1 & 3 & 0 & -4 \\
\hline
  & 1 & 3 & 0 & -4 & 0
\end{array}
$$

$$
\begin{array}{r|rrrr}
-2 & 1 & 3 & 0 & -4 \\
   &   & -2 & -2 & 4 \\
\hline
   & 1 & 1 & -2 & 0
\end{array}
$$

and $x^2 + x - 2 = (x + 2)(x - 1)$.

59. Use synthetic division on the cubic factor in $x(x^3 - 6x^2 + 12x - 8) = 0$.

$$
\begin{array}{r|rrrr}
2 & 1 & -6 & 12 & -8 \\
  &   & 2 & -8 & 8 \\
\hline
  & 1 & -4 & 4 & 0
\end{array}
$$

Since $x^2 - 4x + 4 = (x - 2)^2$, the roots are 2 (with multiplicity 3) and 0.

61. Use synthetic division on the 5th degree factor in $x(x^5 - x^4 - x^3 + x^2 - 12x + 12) = 0$.

$$
\begin{array}{r|rrrrrr}
1 & 1 & -1 & -1 & 1 & -12 & 12 \\
  &   & 1 & 0 & -1 & 0 & -12 \\
\hline
  & 1 & 0 & -1 & 0 & -12 & 0
\end{array}
$$

$$
\begin{array}{r|rrrrr}
2 & 1 & 0 & -1 & 0 & -12 \\
  &   & 2 & 4 & 6 & 12 \\
\hline
  & 1 & 2 & 3 & 6 & 0 \\
\end{array}
$$

$$
\begin{array}{r|rrrr}
-2 & 1 & 2 & 3 & 6 \\
   &   & -2 & 0 & -6 \\
\hline
   & 1 & 0 & 3 & 0 \\
\end{array}
$$

Note, the roots of $x^2 + 3 = 0$ are $\pm i\sqrt{3}$.
Thus, the roots are $x = 0, 1, \pm 2, \pm i\sqrt{3}$.

63. The roots are $\pm 1, -2, 1/4, 3/2$ since

$$
\begin{array}{r|rrrrrr}
-1 & 8 & 2 & -33 & 4 & 25 & -6 \\
   &   & -8 & 6 & 27 & -31 & 6 \\
\hline
   & 8 & -6 & -27 & 31 & -6 & 0 \\
\end{array}
$$

$$
\begin{array}{r|rrrrr}
-2 & 8 & -6 & -27 & 31 & -6 \\
   &   & -16 & 44 & -34 & 6 \\
\hline
   & 8 & -22 & 17 & -3 & 0 \\
\end{array}
$$

$$
\begin{array}{r|rrrr}
1 & 8 & -22 & 17 & -3 \\
  &   & 8 & -14 & 3 \\
\hline
  & 8 & -14 & 3 & 0 \\
\end{array}
$$

and the last quotient is

$$8x^2 - 14x + 4 = (4x - 1)(2x - 3).$$

65. Multiplying by 100, $t^3 - 8t^2 + 11t + 20 = 0$.

$$
\begin{array}{r|rrrr}
4 & 1 & -8 & 11 & 20 \\
  &   & 4 & -16 & -20 \\
\hline
  & 1 & -4 & -5 & 0 \\
\end{array}
$$

Since $t^2 - 4t - 5 = (t - 5)(t + 1) = 0$, we obtain $t = 4$ hr and $t = 5$ hr.

67. Let x be the radius of the cone.
The volume of the cone is $\dfrac{\pi}{3}x^2 \cdot 2$ and the volume of the cylinder with height $4x$ is
$\pi x^2 \cdot (4x)$. So $114\pi = \dfrac{\pi}{3}2x^2 + \pi x^2(4x)$.

Divide by π, multiply by 3, and simplify to obtain $12x^3 + 2x^2 - 342 = 0$.

$$
\begin{array}{r|rrrr}
3 & 12 & 2 & 0 & -342 \\
  &   & 36 & 114 & 342 \\
\hline
  & 12 & 38 & 114 & 0 \\
\end{array}
$$

The radius is $x = 3$ in. since $12x^2 + 38x + 114 = 0$ has no real roots.

For Thought

1. False, since

$$(\sqrt{x - 1} + \sqrt{x})^2 = (x - 1) + 2\sqrt{x(x - 1)} + x.$$

2. False, since -1 is a solution of the first and not of the second equation.

3. False, since -27 is a solution of the first equation but not of the second.

4. False, rather let $u = x^{1/4}$ and $u^2 = x^{1/2}$.

5. True, since $x - 1 = \pm 4^{-3/2}$.

6. False, $\left(-\dfrac{1}{32}\right)^{-2/5} = (-32)^{2/5} = (-2)^2 = 4.$

7. False, $x = -2$ is not a solution.

8. True

9. True

10. False, since $(x^3)^2 = x^6$.

2.5 Exercises

1. Factor: $x^2(x + 3) - 4(x + 3) = 0$
$(x^2 - 4)(x + 3) = (x - 2)(x + 2)(x + 3) = 0$
The solution set is $\{\pm 2, -3\}$.

3. Factor: $2x^2(x + 500) - (x + 500) = 0$
$(2x^2 - 1)(x + 500) = 0$
The solution set is $\left\{\pm\dfrac{\sqrt{2}}{2}, -500\right\}.$

5. Set the right-hand side to 0 and factor.

$$a(a^2 - 15a + 5) = 0$$

$$a = \frac{15 \pm \sqrt{15^2 - 4(1)(5)}}{2} \quad \text{or} \quad a = 0$$

$$a = \frac{15 \pm \sqrt{205}}{2} \quad \text{or} \quad a = 0$$

The solution set is $\left\{ \dfrac{15 \pm \sqrt{205}}{2}, 0 \right\}$.

7. Factor: $3y^2(y^2 - 4) = 3y^2(y - 2)(y + 2) = 0$
The solution set is $\{0, \pm 2\}$.

9. Factor: $(a^2 - 4)(a^2 + 4) =$
$(a - 2)(a + 2)(a - 2i)(a + 2i) = 0$.
The solution set is $\{\pm 2, \pm 2i\}$.

11. Squaring each side, we get

$$x + 1 = x^2 - 10x + 25$$
$$0 = x^2 - 11x + 24 = (x - 8)(x - 3).$$

Checking $x = 3$, we get $2 \neq -2$. Then $x = 3$ is an extraneous root. The solution set is $\{8\}$.

13. Isolate the radical and then square each side.

$$x = x^2 - 40x + 400$$
$$0 = x^2 - 41x + 400 = (x - 25)(x - 16)$$

Checking $x = 16$, we get $2 \neq -6$. Then $x = 16$ is an extraneous root. The solution set is $\{25\}$.

15. Isolate the radical and then square each side.

$$2w = \sqrt{1 - 3w}$$
$$4w^2 = 1 - 3w$$
$$4w^2 + 3w - 1 = (4w - 1)(w + 1) = 0$$
$$w = \frac{1}{4}, -1$$

Checking $w = -1$, we get $-1 \neq 1$.
Then $w = -1$ is an extraneous root.
The solution set is $\left\{ \dfrac{1}{4} \right\}$.

17. Multiply both sides by $z\sqrt{4z + 1}$ and square each side.

$$\sqrt{4z + 1} = 3z$$
$$4z + 1 = 9z^2$$
$$0 = 9z^2 - 4z - 1$$

By the quadratic formula,

$$z = \frac{4 \pm \sqrt{16 - 4(9)(-1)}}{18}$$
$$z = \frac{4 \pm \sqrt{52}}{18} = \frac{4 \pm 2\sqrt{13}}{18} = \frac{2 \pm \sqrt{13}}{9}$$

Since $z = \dfrac{2 - \sqrt{13}}{9} < 0$ and the right-hand side of the original equation is nonnegative,
$z = \dfrac{2 - \sqrt{13}}{9}$ is an extraneous root.

The solution set is $\left\{ \dfrac{2 + \sqrt{13}}{9} \right\}$.

19. Squaring each side, one obtains

$$x^2 - 2x - 15 = 9$$
$$x^2 - 2x - 24 = (x - 6)(x + 4) = 0.$$

The solution set is $\{-4, 6\}$.

21. Isolate a radical and square each side.

$$\sqrt{x + 40} = \sqrt{x} + 4$$
$$x + 40 = x + 8\sqrt{x} + 16$$
$$24 = 8\sqrt{x}$$
$$3 = \sqrt{x}$$
$$9 = x$$

The solution set is $\{9\}$.

23. Isolate a radical and square each side.

$$\sqrt{n + 4} = 5 - \sqrt{n - 1}$$
$$n + 4 = 25 - 10\sqrt{n - 1} + (n - 1)$$
$$-20 = -10\sqrt{n - 1}$$
$$2 = \sqrt{n - 1}$$
$$4 = n - 1$$

The solution set is $\{5\}$.

25. Isolate a radical and square each side.

$$\sqrt{2x + 5} = 9 - \sqrt{x + 6}$$
$$2x + 5 = 81 - 18\sqrt{x + 6} + (x + 6)$$
$$x - 82 = -18\sqrt{x + 6}$$
$$x^2 - 164x + 6724 = 324(x + 6)$$
$$x^2 - 488x + 4780 = 0$$
$$(x - 10)(x - 478) = 0$$

Checking $x = 478$ we get $53 \neq 9$ and $x = 478$ is an extraneous root. The solution set is $\{10\}$.

27. Raise each side to the power 3/2.
Then $x = \pm 2^{3/2} = \pm 8^{1/2} = \pm 2\sqrt{2}$.
The solution set is $\left\{\pm 2\sqrt{2}\right\}$.

29. Raise each side to the power $-3/4$.

Thus, $w = \pm (16)^{-3/4} = \pm (2)^{-3} = \pm \dfrac{1}{8}$.

The solution set is $\left\{\pm \dfrac{1}{8}\right\}$.

31. Raise each side to the power -2.
So, $t = (7)^{-2} = \dfrac{1}{49}$. The solution set is $\left\{\dfrac{1}{49}\right\}$.

33. Raise each side to the power -2.

Then $s - 1 = (2)^{-2} = \dfrac{1}{4}$ and $s = 1 + \dfrac{1}{4}$.

The solution set is $\left\{\dfrac{5}{4}\right\}$.

35. Since $(x^2 - 9)(x^2 - 3) = 0$, the solution set is $\left\{\pm 3, \pm\sqrt{3}\right\}$.

37. Let $u = \dfrac{2c - 3}{5}$ and $u^2 = \left(\dfrac{2c - 3}{5}\right)^2$. Then

$$
\begin{aligned}
u^2 + 2u - 8 &= 0 \\
(u + 4)(u - 2) &= 0 \\
u &= -4, 2 \\
\frac{2c - 3}{5} = -4 \quad &\text{or} \quad \frac{2c - 3}{5} = 2 \\
2c - 3 = -20 \quad &\text{or} \quad 2c - 3 = 10 \\
c = -\frac{17}{2} \quad &\text{or} \quad c = \frac{13}{2}.
\end{aligned}
$$

The solution set is $\left\{-\dfrac{17}{2}, \dfrac{13}{2}\right\}$.

39. Let $u = \dfrac{1}{5x - 1}$ and $u^2 = \left(\dfrac{1}{5x - 1}\right)^2$. Then

$$
\begin{aligned}
u^2 + u - 12 &= (u + 4)(u - 3) = 0 \\
u &= -4, 3 \\
\frac{1}{5x - 1} = -4 \quad &\text{or} \quad \frac{1}{5x - 1} = 3 \\
1 = -20x + 4 \quad &\text{or} \quad 1 = 15x - 3 \\
x = \frac{3}{20} \quad &\text{or} \quad \frac{4}{15} = x.
\end{aligned}
$$

The solution set is $\left\{\dfrac{3}{20}, \dfrac{4}{15}\right\}$.

41. Let $u = v^2 - 4v$ and $u^2 = \left(v^2 - 4v\right)^2$. Then

$$
\begin{aligned}
u^2 - 17u + 60 &= (u - 5)(u - 12) = 0 \\
u &= 5, 12 \\
v^2 - 4v = 5 \quad &\text{or} \quad v^2 - 4v = 12 \\
v^2 - 4v - 5 = 0 \quad &\text{or} \quad v^2 - 4v - 12 = 0 \\
(v - 5)(v + 1) = 0 \quad &\text{or} \quad (v - 6)(v + 2) = 0.
\end{aligned}
$$

The solution set is $\{-1, -2, 5, 6\}$.

43. Factor the left-hand side.

$$
\begin{aligned}
\left(\sqrt{x} - 3\right)\left(\sqrt{x} - 1\right) &= 0 \\
\sqrt{x} &= 3, 1 \\
x &= 9, 1
\end{aligned}
$$

The solution set is $\{1, 9\}$.

45. Factor the left-hand side as
$\left(\sqrt{q} - 4\right)\left(\sqrt{q} - 3\right) = 0$. Then $\sqrt{q} = 3, 4$
and the solution set is $\{9, 16\}$.

47. Set the right-hand side to 0 and factor.

$$
\begin{aligned}
x^{2/3} - 7x^{1/3} + 10 &= 0 \\
\left(x^{1/3} - 5\right)\left(x^{1/3} - 2\right) &= 0 \\
x^{1/3} &= 5, 2
\end{aligned}
$$

The solution set is $\{8, 125\}$.

49. An equivalent statement is

$$
\begin{aligned}
w^2 - 4 = 3 \quad &\text{or} \quad w^2 - 4 = -3 \\
w^2 = 7 \quad &\text{or} \quad w^2 = 1.
\end{aligned}
$$

The solution set is $\left\{\pm\sqrt{7}, \pm 1\right\}$.

51. An equivalent statement assuming $5v \geq 0$ is

$$
\begin{aligned}
v^2 - 3v = 5v \quad &\text{or} \quad v^2 - 3v = -5v \\
v^2 - 8v = 0 \quad &\text{or} \quad v^2 + 2v = 0 \\
v(v - 8) = 0 \quad &\text{or} \quad v(v + 2) = 0 \\
v &= 0, 8, -2.
\end{aligned}
$$

Since $5v \geq 0$, $v = -2$ is an extraneous root and the solution set is $\{0, 8\}$.

53. An equivalent statement is

$$
\begin{aligned}
x^2 - x - 6 = 6 \quad &\text{or} \quad x^2 - x - 6 = -6 \\
x^2 - x - 12 = 0 \quad &\text{or} \quad x^2 - x = 0 \\
(x - 4)(x + 3) = 0 \quad &\text{or} \quad x(x - 1) = 0.
\end{aligned}
$$

The solution set is $\{-3, 0, 1, 4\}$.

55. An equivalent statement is

$$x + 5 = 2x + 1 \quad \text{or} \quad x + 5 = -(2x + 1)$$
$$4 = x \quad \text{or} \quad x = -2.$$

The solution set is $\{-2, 4\}$.

57. Isolate a radical and square both sides.

$$\sqrt{16x + 1} = \sqrt{6x + 13} - 1$$
$$16x + 1 = (6x + 13) - 2\sqrt{6x + 13} + 1$$
$$10x - 13 = -2\sqrt{6x + 13}$$

$$100x^2 - 260x + 169 = 4(6x + 13)$$
$$100x^2 - 284x + 117 = 0$$

$$x = \frac{284 \pm \sqrt{284^2 - 4(100)(117)}}{200}$$
$$x = \frac{284 \pm 184}{200}$$
$$x = \frac{1}{2}, \frac{117}{50}$$

Checking $x = \frac{117}{50}$ we get $\sqrt{\frac{1922}{50}} - \sqrt{\frac{1352}{50}} > 0$

and so $x = \frac{117}{50}$ is an extraneous root.

The solution set is $\left\{\frac{1}{2}\right\}$.

59. Factor as a difference of two squares and then as a sum and difference of two cubes.

$$(v^3 - 8)(v^3 + 8) = 0$$
$$(v - 2)(v^2 + 2v + 4)(v + 2)(v^2 - 2v + 4) = 0$$

Then $v = \pm 2$ or

$$v = \frac{-2 \pm \sqrt{2^2 - 16}}{2} \quad \text{or} \quad v = \frac{2 \pm \sqrt{2^2 - 16}}{2}$$
$$v = \frac{-2 \pm 2i\sqrt{3}}{2} \quad \text{or} \quad v = \frac{2 \pm 2i\sqrt{3}}{2}$$

The solution set is $\left\{\pm 2, -1 \pm i\sqrt{3}, 1 \pm i\sqrt{3}\right\}$.

61. Raise both sides to the power 4. Then

$$7x^2 - 12 = x^4$$
$$0 = x^4 - 7x^2 + 12 = (x^2 - 4)(x^2 - 3)$$
$$x = \pm 2, \pm\sqrt{3}$$

Since the left-hand side of the given equation is nonnegative, $x = -2, -\sqrt{3}$ are extraneous roots. The solution set is $\left\{\sqrt{3}, 2\right\}$.

63. Raise both sides to the power 3.

$$2 + x - 2x^2 = x^3$$
$$x^3 + 2x^2 - x - 2 = 0$$
$$x^2(x + 2) - (x + 2) = (x^2 - 1)(x + 2) = 0$$
$$x = \pm 1, -2$$

The solution set is $\{\pm 1, -2\}$.

65. Let $t = \frac{x - 2}{3}$ and $t^2 = \left(\frac{x - 2}{3}\right)^2$. Then

$$t^2 - 2t + 10 = 0$$
$$t^2 - 2t + 1 = -10 + 1$$
$$(t - 1)^2 = -9$$
$$t = 1 \pm 3i$$
$$\frac{x - 2}{3} = 1 \pm 3i$$
$$x - 2 = 3 \pm 9i$$
$$x = 5 \pm 9i.$$

The solution set is $\{5 \pm 9i\}$.

67. Raise both sides to the power 5/2. Then

$$3u - 1 = \pm 2^{5/2}$$
$$3u - 1 = \pm 32^{1/2}$$
$$3u = 1 \pm 4\sqrt{2}$$

The solution set is $\left\{\frac{1 \pm 4\sqrt{2}}{3}\right\}$.

69. Factor this quadratic type expression.

$$(x^2 + 1) - 11\sqrt{x^2 + 1} + 30 = 0$$
$$\left(\sqrt{x^2 + 1} - 5\right)\left(\sqrt{x^2 + 1} - 6\right) = 0$$
$$\sqrt{x^2 + 1} = 5 \quad \text{or} \quad \sqrt{x^2 + 1} = 6$$
$$x^2 = 24 \quad \text{or} \quad x^2 = 35$$
$$x = \pm 2\sqrt{6}, \pm\sqrt{35}$$

The solution set is $\left\{\pm\sqrt{35}, \pm 2\sqrt{6}\right\}$.

71. An equivalent statement is

$$x^2 - 2x = 3x - 6 \quad \text{or} \quad x^2 - 2x = -3x + 6$$
$$x^2 - 5x + 6 = 0 \quad \text{or} \quad x^2 + x - 6 = 0$$
$$(x - 3)(x - 2) = 0 \quad \text{or} \quad (x + 3)(x - 2) = 0$$
$$x = 2, \pm 3$$

The solution set is $\{2, \pm 3\}$.

73. Raise both sides to the power $-5/3$. Then

$$3m + 1 = \left(-\frac{1}{8}\right)^{-5/3}$$

$$3m + 1 = \left(-\frac{1}{2}\right)^{-5}$$

$$3m + 1 = -32.$$

The solution set is $\{-11\}$.

75. An equivalent statement assuming $x - 2 \geq 0$ is

$$x^2 - 4 = x - 2 \quad \text{or} \quad x^2 - 4 = -x + 2$$
$$x^2 - x - 2 = 0 \quad \text{or} \quad x^2 + x - 6 = 0$$
$$(x - 2)(x + 1) = 0 \quad \text{or} \quad (x + 3)(x - 2) = 0$$
$$x = 2, -1, -3.$$

Since $x - 2 \geq 0$, $x = -1, -3$ are extraneous roots and the solution set is $\{2\}$.

77. Solve for S.

$$21.24 + 1.25S^{1/2} - 9.8(18.34)^{1/3} = 16.296$$
$$1.25S^{1/2} - 25.84396 \approx -4.944$$
$$S^{1/2} \approx 16.72$$
$$S \approx 279.56$$

The maximum sailing area is 279.56 m^2.

79. Let x and $x + 6$ be two numbers. Then

$$\sqrt{x + 6} - \sqrt{x} = 1$$
$$\sqrt{x + 6} = \sqrt{x} + 1$$
$$x + 6 = x + 2\sqrt{x} + 1$$
$$5 = 2\sqrt{x}$$
$$25 = 4x$$
$$x = \frac{25}{4}$$

Since $\frac{25}{4} + 6 = \frac{49}{4}$, the numbers are

$$\frac{25}{4} \text{ and } \frac{49}{4}.$$

81. Solving for d, we find

$$598.9\left(\frac{d}{64}\right)^{-2/3} = 14.26$$

$$\left(\frac{d}{64}\right)^{-2/3} = \frac{14.26}{598.9}$$

$$\frac{d}{64} = \left(\frac{14.26}{598.9}\right)^{-3/2}$$

$$d = 64\left(\frac{14.26}{598.9}\right)^{-3/2}$$

$$d \approx 17,419.3 \text{ lbs.}$$

83. By choosing an appropriate coordinate system, we can assume the circle is given by $(x + r)^2 + (y - r)^2 = r^2$ where $r > 0$ is the radius of the circle and $(-5, 1)$ is the common point between the block and the circle. Note, the radius is less than 5 feet. Substitute $x = -5$ and $y = 1$. Then we obtain

$$(-5 + r)^2 + (1 - r)^2 = r^2$$
$$r^2 - 10r + 25 + 1 - 2r + r^2 = r^2$$
$$r^2 - 12r + 26 = 0.$$

The solutions of the last quadratic equation are $r = 6 \pm \sqrt{10}$. Since $r < 5$, the radius of the circle is $r = 6 - \sqrt{10}$ ft.

85. If Lauren ran 5 miles downstream and swam across for a mile, then this would take her

$$\frac{5}{10} + \frac{1}{8} = 0.625 \text{ hours or } 37.5 \text{ minutes.}$$

The diagonal distance between points A and B is $\sqrt{26}$ miles. Swimming diagonally, it would

take her $\dfrac{\sqrt{26}}{8} \approx 0.6374$ hrs. or 38.2 minutes.

If x is the number of miles she ran, then

$$\frac{x}{10} + \frac{\sqrt{(5 - x)^2 + 1}}{8} = \frac{36}{60}$$
$$\frac{\sqrt{x^2 - 10x + 26}}{8} = \frac{6 - x}{10}$$
$$100(x^2 - 10x + 26) = 64(x^2 - 12x + 36)$$
$$36x^2 - 232x + 296 = 0$$
$$9x^2 - 58x + 74 = 0$$

$$x = \frac{58 \pm \sqrt{(-58)^2 - 4(9)(74)}}{18} \approx 1.75, 4.69$$

To complete the race in 36 minutes, Lauren has to run either 1.75 or 4.69 miles.

For Thought

1. False, to be symmetric about the origin one must have $P(-x) = -P(x)$ for *all* x in the domain.

2. True   **3.**   True

4. False, since $f(-x) = -f(x)$.

5. True   **6.**   True   **7.**   False

8. False, y-intercept is $(0, 38)$.   **9.**   True

10. False, only one x-intercept.

2.6 Exercises

1. y-axis, since $f(-x) = f(x)$

3. $x = 3/2$, since $\dfrac{3}{2}$ is the x-coordinate of the vertex of a parabola

5. None

7. Origin, since $f(-x) = -f(x)$

9. $x = 5$, since 5 is the x-coordinate of the vertex of a parabola

11. Origin, since $f(-x) = -f(x)$

13. It does not cross $(4, 0)$ since $x - 4$ is raised to an even power.

15. It crosses $(1/2, 0)$.

17. It crosses $(1/4, 0)$

19. No x-intercepts since $x^2 - 3x + 10 = 0$ has no real root.

21. It crosses $(3, 0)$ and not $(0, 0)$ since $x^3 - 3x^2 = x^2(x - 3)$.

23. It crosses $(1/2, 0)$ and not $(1, 0)$ since

$$
\begin{array}{r|rrrr}
1 & 2 & -5 & 4 & -1 \\
  &   & 2 & -3 & 1 \\
\hline
  & 2 & -3 & 1 & 0
\end{array}
$$

and $2x^3 - 5x^2 + 4x - 1 = (x - 1)(2x^2 - 3x + 1) =$ $(x - 1)^2(2x - 1)$.

25. It crosses $(2, 0)$ and not $(-3, 0)$ since

$$
\begin{array}{r|rrrr}
-3 & -2 & -8 & 6 & 36 \\
   &   & 6 & 6 & -36 \\
\hline
   & -2 & -2 & 12 & 0
\end{array}
$$

and $(x + 3)(-2x^2 - 2x + 12) =$ $-2(x + 3)(x^2 + x - 6) =$ $-2(x + 3)(x + 3)(x - 2) = -2(x + 3)^2(x - 2)$.

27. $y \to \infty$ **29.** $y \to -\infty$ **31.** $y \to -\infty$

33. $y \to \infty$ **35.** $y \to \infty$

37. Neither symmetry, crosses $(-2, 0)$, does not cross $(1, 0)$, $y \to \infty$ as $x \to \infty$, $y \to -\infty$ as $x \to -\infty$

39. Symmetric about y-axis; no x-intercepts; $y \to \infty$ as $x \to \infty$; $y \to \infty$ as $x \to -\infty$

41. Applying the leading coefficient test, we find

$$\lim_{x \to \infty} (x^2 - 4) = \infty.$$

43. Applying the leading coefficient test, we obtain

$$\lim_{x \to \infty} (-x^5 - x^2) = -\infty.$$

45. Using the leading coefficient test, we get

$$\lim_{x \to -\infty} (-3x) = \infty.$$

47. Using the leading coefficient test, we get

$$\lim_{x \to -\infty} (-2x^2 + 1) = -\infty.$$

49. The graph of $f(x) = (x - 1)^2(x + 3)$ is shown below.

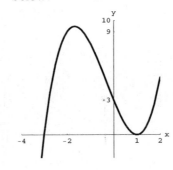

51. The graph of $f(x) = -2(2x-1)^2(x+1)^3$ is given below.

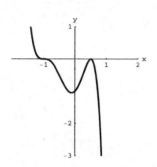

53. e, line

55. g, cubic polynomial and y-intercept $(0,1)$

57. b, y-intercept $(0,6)$, 4th degree polynomial

59. c, y-intercept $(0,-4)$, 3rd degree polynomial

61. $f(x) = x - 30$ has x-intercept $(30,0)$ and y-intercept $(0,-30)$

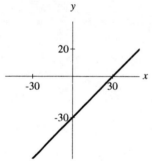

63. $f(x) = (x-30)^2$ does not cross $(30,0)$, y-intercept $(0,900)$, $y \to \infty$ as $x \to \infty$ and as $x \to -\infty$

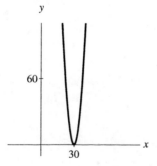

65. $f(x) = x^2(x-40)$ crosses $(40,0)$ but does not cross $(0,0)$, $y \to \infty$ as $x \to \infty$, $y \to -\infty$ as $x \to -\infty$

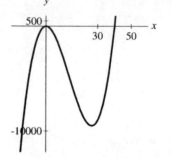

67. $f(x) = (x-20)^2(x+20)^2$ does not cross $(20,0), (-20,0)$, y-intercept $(0,160000)$, $y \to \infty$ as $x \to \infty$ and as $x \to -\infty$

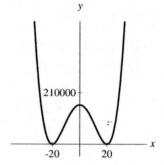

69. Since $f(x) = -x^3 - x^2 + 5x - 3$

1	-1	-1	5	-3
		-1	-2	3
	-1	-2	3	0

$f(x) = (x-1)(-x^2 - 2x + 3) = -(x-1)(x+3)(x-1)$, the graph crosses $(-3,0)$ but not $(1,0)$, y-intercept $(0,-3)$, $y \to -\infty$ as $x \to \infty$, and $y \to \infty$ as $x \to -\infty$

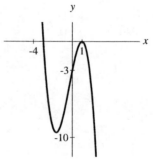

71. Since $x^3 - 10x^2 - 600x = x(x-30)(x+20)$, graph crosses $(0,0), (30,0), (-20,0)$,
$y \to \infty$ as $x \to \infty$, $y \to -\infty$ as $x \to -\infty$

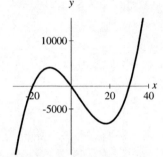

73. $f(x) = x^3 + 18x^2 - 37x + 60$ has only one x-intercept since

$$
\begin{array}{r|rrrr}
-20 & 1 & 18 & -37 & 60 \\
    &   & -20 & 40 & -60 \\
\hline
    & 1 & -2 & 3 & 0
\end{array}
$$

and $x^2 - 2x + 3$ has no real root.
Graph crosses $(-20,0)$, y-intercept $(0,60)$,
$y \to \infty$ as $x \to \infty$, $y \to -\infty$ as $x \to -\infty$

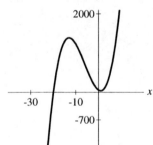

75. Since $-x^2(x^2-196) = -x^2(x-14)(x+14)$, graph crosses $(\pm 14, 0)$ and does not cross $(0,0)$, $y \to -\infty$ as $x \to \infty$ and as $x \to -\infty$

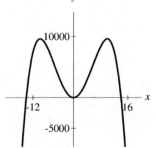

77. Use synthetic division on $f(x) = x^3 + 3x^2 + 3x + 1$ with $c = 1$.

$$
\begin{array}{r|rrrr}
-1 & 1 & 3 & 3 & 1 \\
   &   & -1 & -2 & -1 \\
\hline
   & 1 & 2 & 1 & 0
\end{array}
$$

Since $x^2 + 2x + 1 = (x+1)^2$ then $f(x) = (x+1)^3$. Graph crosses $(-1,0)$, y-intercept $(0,1)$, $y \to \infty$ as $x \to \infty$, and $y \to -\infty$ as $x \to -\infty$.

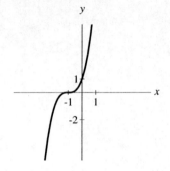

79. The graph crosses $(-7,0)$, but it does not cross $(3,0)$ and $(-5,0)$, y-intercept is $(0,1575)$, $y \to \infty$ as $x \to \infty$, and $y \to -\infty$ as $x \to -\infty$.

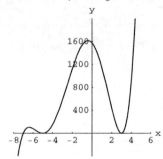

81. The roots of $x(x^2-3) = 0$ are $x = 0, \pm\sqrt{3}$.

If $x = 2$, then $(2)^2 - 3(2) > 0$.
If $x = 1$, then $(1)^2 - 3(1) < 0$.
If $x = -1$, then $(-1)^2 - 3(-1) > 0$.
If $x = -2$, then $(-2)^2 - 3(-2) < 0$.

$$
\begin{array}{ccccccc}
- & 0 & + & 0 & - & 0 & + \\
\end{array}
$$

$$
\xleftarrow{} \quad -2 \quad -\sqrt{3} \quad -1 \quad 0 \quad 1 \quad \sqrt{3} \quad 2 \quad \xrightarrow{}
$$

The solution set is $(-\sqrt{3}, 0) \cup (\sqrt{3}, \infty)$.

83. The roots of $x^2(2 - x^2) = 0$ are $x = 0, \pm\sqrt{2}$.

If $x = 3$, then $2(3)^2 - (3)^4 < 0$.
If $x = 1$, then $2(1)^2 - (1)^4 > 0$.
If $x = -1$, then $2(-1)^2 - (-1)^4 > 0$.
If $x = -3$, then $2(-3)^2 - (-3)^4 < 0$.

```
    −    0    +    0    +    0    −
  <———————————————————————————————>
   −3   −√2   −1    0    -1   √2   3
```

The solution set is $(-\infty, -\sqrt{2}] \cup \{0\} \cup [\sqrt{2}, \infty)$.

85. Let $f(x) = x^3 + 4x^2 - x - 4$. Since

$$x^2(x + 4) - (x + 4) = (x + 4)(x^2 - 1) = 0,$$

the zeros of $f(x)$ are $x = -4, \pm 1$.

If $x = 2$, then $f(2) > 0$.
If $x = 0$, then $f(0) < 0$.
If $x = -2$, then $f(-2) > 0$.
If $x = -5$, then $f(-5) < 0$.

```
    −    0    +    0    −    0    +
  <———————————————————————————————>
   −5   −4   −2   −1    0    1    2
```

The solution set is $(-4, -1) \cup (1, \infty)$.

87. Let $f(x) = x^3 - 4x^2 - 20x + 48$.
Since $f(x) = (x + 4)(x - 2)(x - 6)$, the zeros of $f(x)$ are $x = -4, 2, 6$.

If $x = 7$, then $f(7) > 0$.
If $x = 4$, then $f(4) < 0$.
If $x = 0$, then $f(0) > 0$.
If $x = -5$, then $f(-5) < 0$.

```
    −    0    +    0    −    0    +
  <———————————————————————————————>
   −5   −4    0    2    4    6    7
```

The solution set is $[-4, 2] \cup [6, \infty)$.

89. Note, $f(x) = x^3 - x^2 + x - 1 =$
$x^2(x - 1) + (x - 1) = (x - 1)(x^2 + 1) = 0$
has only one solution, namely, $x = 1$.

If $x = 2$, then $f(2) > 0$.
If $x = 0$, then $f(0) < 0$.

```
        −    0    +
  <———————————————————>
        0    1    2
```

The solution set is $(-\infty, 1)$.

91. Let $f(x) = x^4 - 19x^2 + 90$.
Since $f(x) = (x^2 - 10)(x^2 - 9)$, the zeros of $f(x)$ are $x = \pm\sqrt{10}, \pm 3$.

If $x = 4$, then $f(4) > 0$.
If $x = 3.1$, then $f(3.1) < 0$.
If $x = 0$, then $f(0) > 0$.
If $x = -3.1$, then $f(-3.1) < 0$.
If $x = -4$, then $f(-4) > 0$.

```
   +    0    −    0    +   0   −    0    +
 <—————————————————————————————————————————>
     −√10    −3        3       √10
```

The solution set is $[-\sqrt{10}, -3] \cup [3, \sqrt{10}]$.

93. d, since the x-intercepts of

$$f(x) = \frac{1}{3}(x + 3)(x - 2)$$

are -3 and 2 and $f(0) = -2$

95. The volume function is
$V(x) = x(6 - 2x)(7 - 2x)$ where $0 < x < 3$.
From its graph the maximum possible volume is $V \approx 20.07$ in.^{3}.

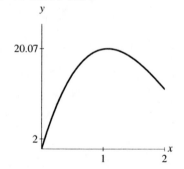

97. Since $3x + 2y = 12$, we get $y = \dfrac{12 - 3x}{2}$.

The volume V of the block is given by

$$
\begin{aligned}
V &= xy\frac{4y}{3} \\
&= \frac{4x}{3}y^2 \\
&= \frac{4x}{3}\left(\frac{12 - 3x}{2}\right)^2 \\
&= 3x^3 - 24x^2 + 48x.
\end{aligned}
$$

One finds from the graph of

$$V = \frac{4x}{3}\left(\frac{12 - 3x}{2}\right)^2$$

that the optimal dimensions of the block are

$$x = \frac{4}{3} \text{ in. by } y = \frac{12 - 4}{2} = 4 \text{ in. by } \frac{16}{3} \text{ in..}$$

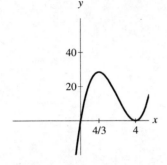

For Thought

1. False, $\sqrt{x} - 3$ is not a polynomial.

2. False, domain is $(-\infty, 2) \cup (2, \infty)$.

3. False

4. False, it has three vertical asymptotes.

5. True

6. False, $y = 5$ is the horizontal asymptote.

7. True   8.   False

9. True, it is an even function.

10. True, since $x = -3$ is not a vertical asymptote.

2.7 Exercises

1. $(-\infty, -2) \cup (-2, \infty)$

3. $(-\infty, -2) \cup (-2, 2) \cup (2, \infty)$

5. $(-\infty, 3) \cup (3, \infty)$

7. $(-\infty, 0) \cup (0, \infty)$

9. $(-\infty, -1) \cup (-1, 0) \cup (0, 1) \cup (1, \infty)$ since

$$f(x) = \frac{3x^2 - 1}{x(x^2 - 1)}$$

11. $(-\infty, -3) \cup (-3, -2) \cup (-2, \infty)$ since

$$f(x) = \frac{-x^2 + x}{(x + 3)(x + 2)}$$

13. Domain $(-\infty, 2) \cup (2, \infty)$, asymptotes $y = 0$ and $x = 2$

15. Domain $(-\infty, 0) \cup (0, \infty)$, asymptotes $y = x$ and $x = 0$

17. Asymptotes $x = 2$, $y = 0$

19. Asymptotes $x = \pm 3$, $y = 0$

21. Asymptotes $x = 1$, $y = 2$

23. Asymptotes $x = 0$, $y = x - 2$ since

$$f(x) = x - 2 + \frac{1}{x}$$

25. Asymptotes $x = -1$, $y = 3x - 3$ since

$$\begin{array}{r} 3x - 3 \\ x + 1 \overline{)3x^2 + 0x + 4} \\ \underline{3x^2 + 3x} \\ -3x + 4 \\ \underline{-3x - 3} \\ 7 \end{array}$$

and $f(x) = 3x - 3 + \dfrac{7}{x + 1}$

27. Asymptotes $x = -2$, $y = -x + 6$ since

$$\begin{array}{r} -x + 6 \\ x + 2 \overline{)-x^2 + 4x + 0} \\ \underline{-x^2 - 2x} \\ 6x + 0 \\ \underline{6x + 12} \\ -12 \end{array}$$

and $f(x) = -x + 6 + \dfrac{-12}{x + 2}$

29. Asymptotes $x = 0$, $y = 0$, no x or y-intercept

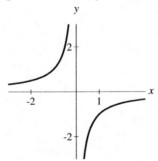

31. Asymptotes $x = 2$, $y = 0$, no x-intercept, y-intercept $(0, -1/2)$

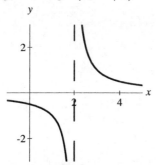

33. Asymptotes $x = \pm 2$, $y = 0$, no x-intercept, y-intercept $(0, -1/4)$

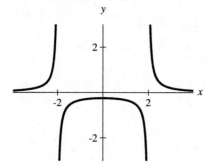

35. Asymptotes $x = -1$, $y = 0$, no x-intercept, y-intercept $(0, -1)$

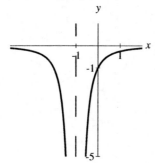

37. Asymptotes $x = 1$, $y = 2$, x-intercept $(-1/2, 0)$, y-intercept $(0, -1)$

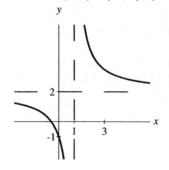

39. Asymptotes $x = -2$, $y = 1$, x-intercept $(3, 0)$, y-intercept $(0, -3/2)$

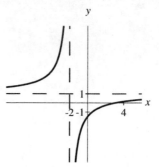

41. Asymptotes $x = \pm 1$, $y = 0$, x- and y-intercept is $(0, 0)$

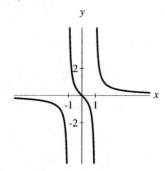

43. Since $f(x) = \dfrac{4x}{(x - 1)^2}$, asymptotes are

$x = 1$, $y = 0$, x- and y-intercept is $(0, 0)$

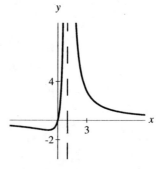

45. Asymptotes are $x = \pm 3$, $y = -1$, x-intercept $(\pm 2\sqrt{2}, 0)$, y-intercept $(0, -8/9)$

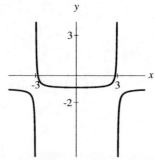

47. Since $f(x) = \dfrac{2x^2 + 8x + 2}{(x+1)^2}$,

asymptotes are $x = -1$, $y = 2$,
by solving $2x^2 + 8x + 2 = 0$ one gets the
x-intercepts $\left(-2 \pm \sqrt{3}, 0\right)$,
y-intercept $(0, 2)$

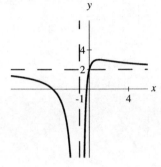

49. We see from the graph that $\lim\limits_{x \to \infty} \dfrac{1}{x^2} = 0$

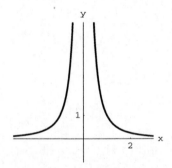

51. From the graph, we find $\lim\limits_{x \to \infty} \dfrac{2x-3}{x-1} = 2$

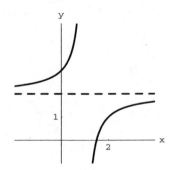

53. From the graph, we find $\lim\limits_{x \to 0^+} \dfrac{1}{x^2} = \infty$

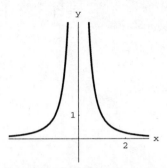

55. We see from the graph that $\lim\limits_{x \to 1^+} \dfrac{2}{x-1} = \infty$

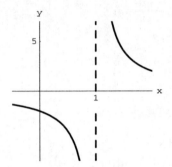

57. Since $f(x) = x + \dfrac{1}{x}$, oblique asymptote is

$y = x$, asymptote $x = 0$, no x-intercept,
no y-intercept, graph goes through
$(1, 2), (-1, -2)$

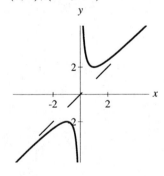

59. Since $f(x) = x - \dfrac{1}{x^2}$, oblique asymptote

is $y = x$, asymptote $x = 0$, x-intercept $(1, 0)$,
no y-intercept, graph goes through $(-1, -2)$

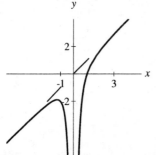

61. $f(x) = x - 1 + \dfrac{1}{x+1}$ since

$$
\begin{array}{r}
x - 1 \\
x + 1 \;\overline{)x^2 + 0x} \\
\underline{x^2 + x} \\
-x + 0 \\
\underline{-x - 1} \\
1
\end{array}
$$

Oblique asymptote $y = x - 1$,
asymptote $x = -1$, x-intercept $(0, 0)$,
graph goes through $(-2, -4)$

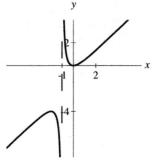

63. $f(x) = 2x + 1 + \dfrac{1}{x-1}$ since

$$
\begin{array}{r}
2x + 1 \\
x - 1 \;\overline{)2x^2 - x + 0} \\
\underline{2x^2 - 2x} \\
x + 0 \\
\underline{x - 1} \\
1
\end{array}
$$

Oblique asymptote $y = 2x + 1$,
asymptote $x = 1$,
x-intercepts $(0, 0), (1/2, 0)$,
graph goes through $(2, 6)$

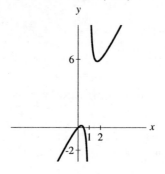

65. Using long division, we find

$$
\frac{x^3 - x^2 - 4x + 5}{x^2 - 4} = (x - 1) + \frac{1}{x^2 - 4}.
$$

The oblique asymptote is $y = x - 1$.

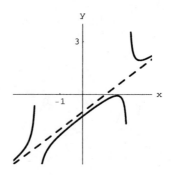

67. Using long division, we find

$$\frac{-x^3 + x^2 + 5x - 4}{x^2 + x - 2} = (-x + 2) + \frac{x}{x^2 + x - 2}.$$

The oblique asymptote is $y = -x + 2$.

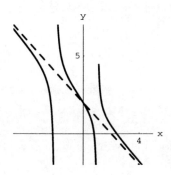

69. e **71.** a

73. b **75.** c

77. Domain is $\{x : x \neq \pm 1\}$ since

$$f(x) = \frac{x+1}{(x+1)(x-1)} = \frac{1}{x-1} \text{ if } x \neq -1,$$

a 'hole' at $(-1, -1/2)$, asymptotes $x = 1$, $y = 0$, no x-intercept, y-intercept $(0, -1)$

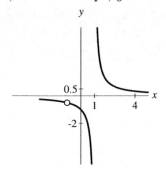

79. The domain is $\{x : x \neq 1\}$ since

$$f(x) = \frac{(x-1)(x+1)}{x-1} = x + 1 \text{ where } x \neq 1,$$

a line with a 'hole' at $(1, 2)$

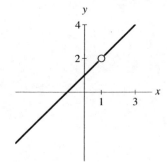

81. The sign graph of $\dfrac{x-4}{x+2} \leq 0$ is

```
- - - - - - - - 0 + + + +
- - - - 0 + + + + + + + +
<———————————————————————>
     -2        4
```

The solution set is $(-2, 4]$.

83. The sign graph of $\dfrac{q+8}{q+3} > 0$ is

```
- - - - - - - - 0 + + + +
- - - - 0 + + + + + + + +
<———————————————————————>
     -8        -3
```

The solution set is $(-\infty, -8) \cup (-3, \infty)$.

85. The sign graph of $\dfrac{(w-3)(w+2)}{w-6} \geq 0$ is

```
- - - - - - - - - - - - 0 + + +
- - - - - - - - - 0 + + + + + +
- - - - - 0 + + + + + + + + + +
<———————————————————————————————>
    -2        3        6
```

The solution set is $[-2, 3] \cup (6, \infty)$.

87. The sign graph of $\dfrac{-5}{(x+2)(x-3)} > 0$ is

```
- - - - - - - - - - - - - - - - - -
- - - - - - - - - 0 + + + + + +
- - - - - - - 0 + + + + + + + + + +
<———————————————————————————————>
     -2        3
```

The solution set is $(-2, 3)$.

89. The sign graph of $\dfrac{(x-4)(x+2)}{5-x} > 0$ is

```
+ + + + + + + + + + + + + + + + 0 - - -
- - - - - - - - - - 0 + + + + + +
- - - - - 0 + + + + + + + + + +
<———————————————————————————————>
    -2        4        5
```

The solution set is $(-\infty, -2) \cup (4, 5)$.

91. Let $R(x) = \dfrac{(x-3)(x+1)}{x-5} \geq 0.$

Since $R(-2) < 0, R(0) > 0, R(4) < 0,$
and $R(7) > 0$, we obtain

$$
\begin{array}{ccccccc}
- & 0 & + & 0 & - & \text{U} & + \\
\end{array}
$$

$$
\begin{array}{ccccccc}
-2 & -1 & 0 & 3 & 4 & 5 & 7 \\
\end{array}
$$

The solution set is $[-1,3] \cup (5,\infty)$.

93. Let $f(x) = \dfrac{(x-\sqrt{7})(x+\sqrt{7})}{(\sqrt{2}-x)(\sqrt{2}+x)}.$

If $x = 3$, then $f(3) < 0$.
If $x = 2$, then $f(2) > 0$.
If $x = 0$, then $f(0) < 0$.
If $x = -2$, then $f(-2) > 0$.
If $x = -3$, then $f(-3) < 0$.

$$
\begin{array}{ccccccc}
- & 0 & + & \text{U} & - & \text{U} & + & 0 & - \\
\end{array}
$$

$$
\begin{array}{cccc}
-\sqrt{7} & -\sqrt{2} & \sqrt{2} & \sqrt{7} \\
\end{array}
$$

The solution set is

$$(-\infty, -\sqrt{7}] \cup (-\sqrt{2}, \sqrt{2}) \cup [\sqrt{7}, \infty).$$

95. Let $R(x) = \dfrac{(x+1)^2}{(x-5)(x+3)} \geq 0.$

Since $R(-4) > 0, R(-2) < 0, R(0) < 0,$
and $R(6) > 0$, we get

$$
\begin{array}{ccccccc}
+ & \text{U} & - & 0 & - & \text{U} & + \\
\end{array}
$$

$$
\begin{array}{ccccccc}
-4 & -3 & -2 & -1 & 0 & 5 & 6 \\
\end{array}
$$

The solution set is $(-\infty, -3) \cup \{-1\} \cup (5, \infty)$.

97. Since the sign graph of $\dfrac{w-1}{w^2} > 0$ is

$$
\begin{array}{c}
- - - - - - - - 0 + + + + \\
+ + + + 0 + + + + + + + +
\end{array}
$$

$$
\begin{array}{cc}
0 & 1 \\
\end{array}
$$

then the solution set is the interval $(1, \infty)$.

99. Since the sign graph of $\dfrac{w^2 - 4w + 5}{w-3} > 0$ is

$$
\begin{array}{c}
- - - - - 0 + + + + + + + + + \\
\end{array}
$$

$$
3
$$

the solution is the interval $(3, \infty)$.

101. Average cost is $C = \dfrac{100 + x}{x}$. If she visits the

zoo 100 times, then her average cost per visit

is $C = \dfrac{200}{100} = \$2$. Since $C \to 1$ as $x \to \infty$,

over a long period her average cost per visit is $\$1$.

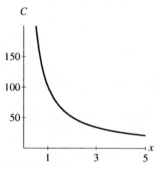

103. Since the second half is 100 miles long and it must be completed in $4 - x$ hours, the average speed for the second half is $S(x) = \dfrac{100}{4-x}$. The asymptote $x = 4$ implies the average speed in the second half goes to ∞ as the completion time in the first half shortens to four hours.

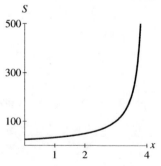

105.

(a) A graph of $C = 400x + \dfrac{10,000}{x}$ is given

on the next page.

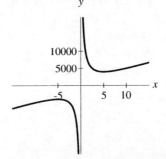

(b) From the graph above, C is minimized when $x = 5$.

Chapter 2 Review Exercises

1. By using the method of completing the square, we obtain

$$
\begin{aligned}
f(x) &= 3\left(x^2 - \frac{2}{3}x + \frac{1}{9}\right) - \frac{1}{3} + 1 \\
&= 3\left(x - \frac{1}{3}\right)^2 + \frac{2}{3}
\end{aligned}
$$

3. Since $y = 2(x^2 - 2x + 1) - 2 - 1 = 2(x-1)^2 - 3$, the vertex is $(1, -3)$ and axis of symmetry is $x = 1$. Setting $y = 0$, we get $x - 1 = \pm\dfrac{\sqrt{6}}{2}$ and the x-intercepts are $\left(\dfrac{2 \pm \sqrt{6}}{2}, 0\right)$.

The y-intercept is $(0, -1)$.

5. From the x-intercepts, $y = a(x+1)(x-3)$. Substitute $(0, 6)$, so $6 = a(-3)$ or $a = -2$. The equation of the parabola is $y = -2(x^2 - 2x - 3)$ or $y = -2x^2 + 4x + 6$.

7. $-1 - i$

9. $16 - 40i + 25i^2 = 16 - 25 - 40i = -9 - 40i$

11. $2 + 6i - 6i - 18i^2 = 20$

13.
$$
\frac{2}{i} - \frac{3i}{i} = \frac{2(-i)}{i(-i)} - 3 = \frac{-2i}{1} - 3 = -3 - 2i
$$

15.
$$
\frac{1-i}{2+i} \cdot \frac{2-i}{2-i} = \frac{1-3i}{5} = \frac{1}{5} - \frac{3}{5}i
$$

17.
$$
\frac{6 + 2i\sqrt{2}}{2} = 3 + i\sqrt{2}
$$

19. $\left(i^4\right)^8 \cdot i^2 + \left(i^4\right)^4 \cdot i^3 = 1 \cdot (-1) + 1 \cdot (-i) = -1 - i$

21 $1/3$

23. $\pm 2\sqrt{2}$

25. Factoring, we obtain
$$
m(x) = (2x - 1)(4x^2 + 2x + 1).
$$

By using the quadratic formula, we obtain the zeros to the second factor. Namely,
$$
x = \frac{-2 \pm \sqrt{-12}}{8} = \frac{-2 \pm 2i\sqrt{3}}{8}.
$$

The zeros are $\dfrac{1}{2}, \dfrac{-1 \pm i\sqrt{3}}{4}$.

27. Since $P(t) = (t^2 - 10)(t^2 + 10)$, the zeros are $\pm\sqrt{10}, \pm i\sqrt{10}$

29. Factoring, we obtain
$$
R(s) = 4s^2(2s-1) - (2s-1) = (4s^2-1)(2s-1).
$$
The zeros are $\dfrac{1}{2}$ (with multiplicity 2) and $-\dfrac{1}{2}$.

31. Find the roots of the second factor of $f(x) = x(x^2 + 2x - 6)$. Since $x^2 + 2x - 6 = (x^2 + 2x + 1) - 6 - 1 = (x+1)^2 - 7$, the zeros are $0, -1 \pm \sqrt{7}$.

33. $P(3) = 108 - 27 + 3 - 1 = 83$. By synthetic division, we get

$$
\begin{array}{r|rrrr}
3 & 4 & -3 & 1 & -1 \\
  &   & 12 & 27 & 84 \\
\hline
  & 4 & 9 & 28 & 83
\end{array}
$$

The remainder is $P(3) = 83$.

35. $P(-1/2) = \dfrac{1}{4} - \dfrac{1}{4} + 3 + 2 = 5.$

By synthetic division, we obtain

$$
\begin{array}{r|rrrrrr}
-1/2 & -8 & 0 & 2 & 0 & -6 & 2 \\
     &   & 4 & -2 & 0 & 0 & 3 \\
\hline
     & -8 & 4 & 0 & 0 & -6 & 5
\end{array}
$$

So remainder is $P(-1/2) = 5.$

37. $\pm \left\{ 1, \dfrac{1}{3}, 2, \dfrac{2}{3} \right\}$

39. $\pm \left\{ 1, \dfrac{1}{2}, \dfrac{1}{3}, \dfrac{1}{6}, 3, \dfrac{3}{2} \right\}$

41.

$$
\begin{aligned}
2 \left(x + \dfrac{1}{2} \right) (x - 3) &= (2x + 1)(x - 3) \\
&= 2x^2 - 5x - 3
\end{aligned}
$$

An equation is $2x^2 - 5x - 3 = 0.$

43. $(x - (3 - 2i)) \, (x - (3 + 2i)) =$

$$
\begin{aligned}
&= ((x - 3) + 2i) \, ((x - 3) - 2i) \\
&= (x - 3)^2 + 4 \\
&= x^2 - 6x + 13
\end{aligned}
$$

An equation is $x^2 - 6x + 13 = 0.$

45. $(x - 2) \, (x - (1 - 2i)) \, (x - (1 + 2i)) =$

$$
\begin{aligned}
&= (x - 2) \, ((x - 1) + 2i) \, ((x - 1) - 2i) \\
&= (x - 2) \left((x - 1)^2 + 4 \right) \\
&= (x - 2)(x^2 - 2x + 5) \\
&= x^3 - 4x^2 + 9x - 10
\end{aligned}
$$

Thus, an equation is $x^3 - 4x^2 + 9x - 10 = 0.$

47. $\left(x - (2 - \sqrt{3}) \right) \left(x - (2 + \sqrt{3}) \right) =$

$$
\begin{aligned}
&= \left((x - 2) + \sqrt{3} \right) \left((x - 2) - \sqrt{3} \right) \\
&= \left((x - 2)^2 - 3 \right) \\
&= x^2 - 4x + 1
\end{aligned}
$$

Thus, an equation is $x^2 - 4x + 1 = 0.$

49. $P(x) = P(-x) = x^8 + x^6 + 2x^2$ has no sign variation. There are 6 imaginary roots and 0 has multiplicity 2.

51. $P(x) = 4x^3 - 3x^2 + 2x - 9$ has 3 sign variations and $P(-x) = -4x^3 - 3x^2 - 2x - 9$ has no sign variation. There are
(a) 3 positive roots, or
(b) 1 positive and 2 imaginary roots.

53. $P(x) = x^3 + 2x^2 + 2x + 1$ has no sign variation and $P(-x) = -x^3 + 2x^2 - 2x + 1$ has 3 sign variations. There are
(a) 3 negative roots, or
(b) 1 negative root and 2 imaginary roots.

55. Roots are $1, 2, 3$ since

$$
\begin{array}{r|rrrr}
1 & 1 & -6 & 11 & -6 \\
  &   & 1 & -5 & 6 \\
\hline
  & 1 & -5 & 6 & 0
\end{array}
$$

and $x^2 - 5x + 6 = (x - 3)(x - 2).$

57. Roots are $\pm i, 1/3, 1/2$ since

$$
\begin{array}{r|rrrrr}
1/2 & 6 & -5 & 7 & -5 & 1 \\
    &   & 3 & -1 & 3 & -1 \\
\hline
    & 6 & -2 & 6 & -2 & 0
\end{array}
$$

$$
\begin{array}{r|rrrr}
1/3 & 6 & -2 & 6 & -2 \\
    &   & 2 & 0 & 2 \\
\hline
    & 6 & 0 & 6 & 0
\end{array}
$$

and the zeros of $6x^2 + 6 = 6(x^2 + 1) = 0$ are $\pm i.$

59. Roots are $3, 3 \pm i$ since

$$
\begin{array}{r|rrrr}
3 & 1 & -9 & 28 & -30 \\
  &   & 3 & -18 & 30 \\
\hline
  & 1 & -6 & 10 & 0
\end{array}
$$

and the zeros of $x^2 - 6x + 10 = (x - 3)^2 + 1 = 0$ are $3 \pm i.$

61. Roots are $2, 1 \pm i\sqrt{2}$ since

$$
\begin{array}{c|cccc}
2 & 1 & -4 & 7 & -6 \\
  &   & 2 & -4 & 6 \\
\hline
  & 1 & -2 & 3 & 0
\end{array}
$$

and the zeros of $x^2 - 2x + 3 = (x-1)^2 + 2 = 0$ are $1 \pm i\sqrt{2}$.

63. Apply synthetic division to the second factor in $x(2x^3 - 5x^2 - 2x + 2) = 0$.

$$
\begin{array}{c|cccc}
1/2 & 2 & -5 & -2 & 2 \\
    &   & 1 & -2 & -2 \\
\hline
    & 2 & -4 & -4 & 0
\end{array}
$$

By completing the square, the zeros of $2x^2 - 4x - 4 = 2(x^2 - 2x) - 4 = 2(x-1)^2 - 6 = 0$ are $1 \pm \sqrt{3}$. All the roots are $x = 0, 1/2, 1 \pm \sqrt{3}$.

65. Solve an equivalent statement assuming $3v \geq 0$.

$$
\begin{array}{rcl}
2v - 1 = 3v & \text{or} & 2v - 1 = -3v \\
-1 = v & \text{or} & 5v = 1
\end{array}
$$

Since $3v \geq 0$, $v = -1$ is an extraneous root. The solution set is $\{1/5\}$.

67. Let $w = x^2$ and $w^2 = x^4$.

$$
\begin{aligned}
w^2 + 7w &= 18 \\
(w+9)(w-2) &= 0 \\
w &= -9, 2 \\
x^2 = -9 \quad \text{or} \quad x^2 &= 2.
\end{aligned}
$$

Since $x^2 = -9$ has no real solution, the solution set is $\{\pm\sqrt{2}\}$.

69. Isolate a radical and square both sides.

$$
\begin{aligned}
\sqrt{x+6} &= \sqrt{x-5} + 1 \\
x + 6 &= x - 5 + 2\sqrt{x-5} + 1 \\
5 &= \sqrt{x-5} \\
25 &= x - 5
\end{aligned}
$$

The solution set is $\{30\}$.

71. Let $w = \sqrt[4]{y}$ and $w^2 = \sqrt{y}$.

$$
\begin{aligned}
w^2 + w - 6 &= 0 \\
(w+3)(w-2) &= 0 \\
w = -3 \quad &\text{or} \quad w = 2 \\
y^{1/4} = -3 \quad &\text{or} \quad y^{1/4} = 2
\end{aligned}
$$

Since $y^{1/4} = -3$ has no real solution, the solution set is $\{16\}$.

73. Let $w = x^2$ and $w^2 = x^4$.

$$
\begin{aligned}
w^2 - 3w - 4 &= 0 \\
(w+1)(w-4) &= 0 \\
x^2 = -1 \quad &\text{or} \quad x^2 = 4
\end{aligned}
$$

Since $x^2 = -1$ has no real solution, the solution set is $\{\pm 2\}$.

75. Raise to the power $3/2$ and get $x - 1 = \pm(4^{1/2})^3$. So, $x = 1 \pm 8$. The solution set is $\{-7, 9\}$.

77. No solution since $(x+3)^{-3/4}$ is nonnegative.

79. Since $3x - 7 = 4 - x$, we obtain $4x = 11$. The solution set is $\{11/4\}$.

81. Symmetric about $x = 3/4$ since $\dfrac{-b}{2a} = \dfrac{3}{4}$.

83. Symmetric about y-axis since $f(-x) = f(x)$.

85. Symmetric about the origin for $f(-x) = -f(x)$.

87. $(-\infty, -2.5) \cup (-2.5, \infty)$

89. $(-\infty, \infty)$

91. Since $f(x) = (x-2)(x+1)$, the x-intercepts are $(2,0), (-1,0)$, y-intercept is $(0,-2)$. Since $\dfrac{-b}{2a} = \dfrac{1}{2}$, the vertex is $(1/2, -9/4)$.

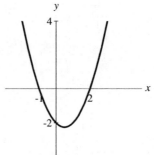

93. Use synthetic division on $f(x) = x^3 - 3x - 2$.

$$\begin{array}{c|cccc} -1 & 1 & 0 & -3 & -2 \\ & & -1 & 1 & 2 \\ \hline & 1 & -1 & -2 & 0 \end{array}$$

Since $x^2 - x - 2 = (x - 2)(x + 1)$,
$f(x) = (x + 1)^2(x - 2)$. Graph crosses $(2, 0)$
but does not cross $(-1, 0)$. y-intercept is
$(0, -2)$.

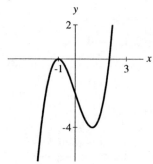

95. Use synthetic division on

$$f(x) = \frac{1}{2}x^3 - \frac{1}{2}x^2 - 2x + 2.$$

$$\begin{array}{c|cccc} 1 & 1/2 & -1/2 & -2 & 2 \\ & & 1/2 & 0 & -2 \\ \hline & 1/2 & 0 & -2 & 0 \end{array}$$

Since the roots of $\frac{1}{2}x^2 - 2 = 0$ are ± 2,
the graph crosses x-intercepts $(\pm 2, 0), (1, 0)$,
y-intercept is $(0, 2)$.

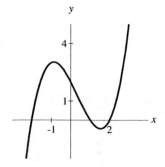

97. Factoring, we get $f(x) = \frac{1}{4}(x^4 - 8x^2 + 16) =$
$\frac{1}{4}(x^2 - 4)^2$. The graph does not cross through
x-intercepts $(\pm 2, 0)$, y-intercept is $(0, 4)$.

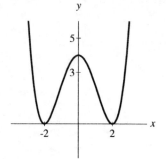

99. $f(x) = \dfrac{2}{x + 3}$ has no x-intercept, y-intercept
is $(0, 2/3)$, asymptotes are $x = -3$, $y = 0$

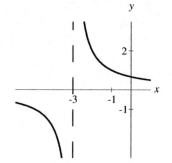

101. $f(x) = \dfrac{2x}{x^2 - 4}$ has x-intercept $(0, 0)$,
asymptotes are $x = \pm 2$, $y = 0$.
Symmetric about the origin.
Graph goes through $(1, -2/3)$, $(3, 6/5)$,
and $(-3, -6/5)$.

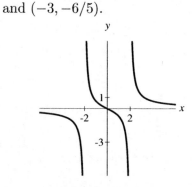

103. $f(x) = \dfrac{(x-1)^2}{x-2}$ has x-intercept $(1,0)$,

y-intercept is $(0, -1/2)$, asymptote $x = 2$,
and oblique asymptote $y = x$ since

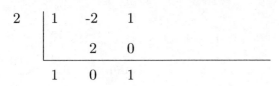

and $f(x) = x + \dfrac{1}{x-2}$.

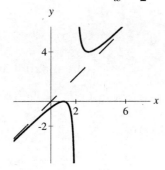

105. $f(x) = \dfrac{2x-1}{2-x}$ has x-intercept $(1/2, 0)$,

y-intercept is $(0, -1/2)$,
asymptotes are $x = 2$ and $y = -2$,
graph goes through $(1,1), (3,-5)$.

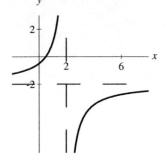

107. Since $f(x) = \dfrac{x^2-4}{x-2} = x + 2$ if $x \neq 2$,

x-intercept is $(-2, 0)$, y-intercept is $(0, 2)$,
no asymptotes

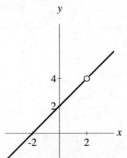

109. Set the right side to zero to obtain

$$8x^2 - 6x + 1 < 0.$$

Factoring, we get $(4x-1)(2x-1) < 0$.
The sign graph is shown on the next page.

```
- - - - - - - - - 0 + + + +
- - - - 0 + + + + + + + +
<--------------------------->
        1/4         1/2
```

The solution set is the interval $(1/4, 1/2)$.

111. The sign graph of $(3-x)(x+5) \geq 0$ is shown
below.

```
+ + + + + + + + 0 - - - -
- - - - 0 + + + + + + + +
<--------------------------->
       -5           3
```

The solution set is the interval $[-5, 3]$.

113. Using the Rational Zero Theorem and synthetic division, we obtain

$$
\begin{array}{r|rrrr}
1/2 & 4 & -400 & -1 & 100 \\
    &   & 2 & -199 & -100 \\
\hline
    & 4 & -398 & -200 & 0
\end{array}
$$

Note, $4x^2 - 398x - 200 = (4x+2)(x-100)$. Using the zeros, we obtain the sign graph of $4x^3 - 400x^2 - x + 100 \geq 0$. Namely,

```
- - - - - - - - - - - - - - - 0 + + +
- - - - - - - - - - 0 + + + + + +
- - - - - 0 + + + + + + + + + +
    ←————————————————————→
        -1/2    1/2      100
```

The solution set is $\left[-\dfrac{1}{2}, \dfrac{1}{2}\right] \cup [100, \infty)$.

115. Let $R(x) = \dfrac{x+10}{x+2} - 5 = \dfrac{-4x}{x+2} < 0.$

Using the test-point method, we find $R(-3) < 0$, $R(-1) > 0$, and $R(1) < 0$.

```
        -    U    +    0    -
    ←————————————————————→
        -3  -2   -1    0    1
```

The solution is $(-\infty, -2) \cup (0, \infty)$.

117. $R(x) = \dfrac{12 - 7x}{x^2} + 1 = \dfrac{(x-3)(x-4)}{x^2} > 0.$

Using the test-point method, we get $R(-1) > 0$, $R(1) > 0$, $R(3.5) < 0$, and $R(5) > 0$.

```
        +    U    +    0    -    0    +
    ←————————————————————————→
       -1    0    1    3   3.5   4    5
```

the solution is $(-\infty, 0) \cup (0, 3) \cup (4, \infty)$.

119. Let $R(x) = \dfrac{(x-1)(x-2)}{(x-3)(x-4)}$. We will use

the test-point method. Note, $R(0) > 0$, $R(1.5) < 0$, $R(2.5) > 0$, $R(3.5) < 0$, and $R(5) > 0$.

```
    +    0    -    0    +    U    -    U    +
←————————————————————————————————————→
    0    1   1.5   2   2.5   3   3.5   4    5
```

The solution is $(-\infty, 1] \cup [2, 3) \cup (4, \infty)$.

121. Quotient $x^2 - 3x$, remainder -15

$$
\begin{array}{r|rrrr}
3 & 1 & -6 & 9 & -15 \\
  &   & 3 & -9 & 0 \\
\hline
  & 1 & -3 & 0 & -15
\end{array}
$$

123. Since $\dfrac{-b}{2a} = \dfrac{-156}{-32} = 4.875$, the maximum

height is $-16(4.875)^2 + 156(4.875) = 380.25$ ft.

125. Since $b = a^2 - 16$, the area A of the triangle is

$$
\begin{aligned}
A &= \frac{1}{2}(\text{base})(\text{height}) \\
A &= \frac{1}{2}(2a)(-b) \\
A &= -ab \\
A &= -a(a^2 - 16) \\
A &= -a^3 + 16a.
\end{aligned}
$$

Note, A is maximized when $a = \dfrac{4}{\sqrt{3}} \approx 2.3$ as

can be seen from the graph of $A = -a(a^2 - 16)$.

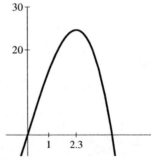

The point $(a, b) = \left(\dfrac{4}{\sqrt{3}}, -\dfrac{32}{3}\right) \approx (2.3, -10.7)$

maximizes the area.

For Thought

1. True   **2.** False

3. False, $5°$ is coterminal with $-355°$.

4. False   **5.** True, since $\dfrac{38\pi}{4} = \dfrac{19\pi}{2}$.

6. False, $210° = \dfrac{7\pi}{6}$.

7. False, since $25°20'40'' \neq 25.34°$; $25.34°$ is an approximation to $25°20'40''$.

8. True, since $2\pi + \dfrac{\pi}{3} = \dfrac{6\pi}{3} + \dfrac{\pi}{3} = \dfrac{7\pi}{3}$.

9. True, since the arc length is $s = \alpha r$ we obtain

$$s = \frac{\pi}{3} \cdot 3 \text{ ft} = \pi \text{ ft}.$$

10. True, since if we substitute $\alpha = 1$ rad into the arc length formula $s = \alpha r$ we get

$$s = 1 \text{ rad} \cdot r = r.$$

3.1 Exercises

1. Substitute $k = 1, 2, -1, -2$ into $60° + k \cdot 360°$ to obtain the coterminal angles

$$420°, 780°, -300°, -660°.$$

There are other coterminal angles.

3. Substitute $k = 1, 2, -1, -2$ into $-16° + k \cdot 360°$ to find the coterminal angles

$$344°, 704°, -376°, -736°.$$

There are other coterminal angles.

5. Yes, since $123.4° - (-236.6°) = 360°$ is an integral multiple of $360°$.

7. No, since $1055° - (155°) = 900° = k \cdot 360°$ does not have an integral solution for any k.

9. Quadrant I

11. $-125°$ lies in Quadrant III since $-125° + 360° = 235°$ and $180° < 235° < 270°$

13. Quadrant IV

15. $750°$ lies in Quadrant I since $750° - 720° = 30°$

17. $45°$

19. $60°$

21. $120°$

23. $400° - 360° = 40°$

25. $-340° + 360° = 20°$

27. $-1100° + 4 \cdot 360° = 340°$

29. $13° + \dfrac{12°}{60} = 13.2°$

31. $-8° - \dfrac{30°}{60} - \dfrac{18}{3600}^{\circ} = -8.505°$

33. $28° + \dfrac{5}{60}^{\circ} + \dfrac{9}{3600}^{\circ} \approx 28.0858°$

35. $75.5° = 75°30'$ since $0.5(60) = 30$

37. $-17.33° = -17°19'48''$ since $0.33(60) = 19.8$ and $0.8(60) = 48$

39. $18.123° \approx 18°7'23''$ since $0.123(60) = 7.38$ and $0.38(60) \approx 23$

41. $\dfrac{\pi}{6}$

43. $18° \cdot \dfrac{\pi}{180} = \dfrac{\pi}{10}$

45. $-67.5° \cdot \dfrac{\pi}{180} = -\dfrac{135\pi}{360} = -\dfrac{3\pi}{8}$

47. $630° \cdot \dfrac{\pi}{180} = \dfrac{7\pi}{2}$

49. $37.4° \cdot \dfrac{\pi}{180} \approx 0.653$

51. $\left(-13 - \dfrac{47}{60}\right) \cdot \dfrac{\pi}{180} \approx -0.241$

53. $-\left(53 + \dfrac{37}{60} + \dfrac{6}{3600}\right) \cdot \dfrac{\pi}{180} \approx -0.936$

55. $\dfrac{5\pi}{12} \cdot \dfrac{180}{\pi} = 75°$

57. $\dfrac{7\pi}{4} \cdot \dfrac{180}{\pi} = 315°$

59. $-6\pi \cdot \dfrac{180}{\pi} = -1080°$

61. $2.39 \cdot \dfrac{180}{\pi} \approx 136.937°$

63. Substitute $k = 1, 2, -1, -2$ into $\dfrac{\pi}{3} + k \cdot 2\pi$ to obtain the coterminal angles

$$\dfrac{7\pi}{3}, \dfrac{13\pi}{3}, -\dfrac{5\pi}{3}, -\dfrac{11\pi}{3}.$$

There are other coterminal angles.

65. Substitute $k = 1, 2, -1, -2$ into $-\dfrac{\pi}{6} + k \cdot 2\pi$ to find the coterminal angles

$$\dfrac{11\pi}{6}, \dfrac{23\pi}{6}, -\dfrac{13\pi}{6}, -\dfrac{25\pi}{6}.$$

There are other coterminal angles.

67. $3\pi - 2\pi = \pi$

69. $\dfrac{9\pi}{2} - 4\pi = \dfrac{\pi}{2}$

71. $-\dfrac{5\pi}{3} + 2\pi = \dfrac{\pi}{3}$

73. $-\dfrac{13\pi}{3} + 6\pi = \dfrac{5\pi}{3}$

75. $8.32 - 2\pi \approx 2.04$

77. No, since $\dfrac{29\pi}{4} - \dfrac{3\pi}{4} = \dfrac{26\pi}{4} = k \cdot 2\pi$ does not have an integral solution for any k.

79. Yes, since $\dfrac{7\pi}{6} - \dfrac{-5\pi}{6} = \dfrac{12\pi}{6} = 2\pi$.

81. Quadrant I

83. Quadrant III

85. $\dfrac{13\pi}{8}$ lies in Quadrant IV since

$$\dfrac{3\pi}{2} = \dfrac{12\pi}{8} < \dfrac{13\pi}{8} < 2\pi$$

87. Note $2\pi \approx 6.28$ and $3\pi/2 \approx 4.71$. Since -7.3 is coterminal with $-7.3 + 2(6.28) = 5.26$ and $4.71 < 5.26 < 6.28$, it follows that -7.3 lies in Quadrant IV.

89. $30° = \dfrac{\pi}{6}$, $45° = \dfrac{\pi}{4}$, $60° = \dfrac{\pi}{3}$, $90° = \dfrac{\pi}{2}$,

$120° = \dfrac{2\pi}{3}$, $135° = \dfrac{3\pi}{4}$, $150° = \dfrac{5\pi}{6}$, $180° = \pi$,

$210° = \dfrac{7\pi}{6}$, $225° = \dfrac{5\pi}{4}$, $240° = \dfrac{4\pi}{3}$,

$270° = \dfrac{3\pi}{2}$, $300° = \dfrac{5\pi}{3}$, $315° = \dfrac{7\pi}{4}$,

$330° = \dfrac{11\pi}{6}$, $360° = 2\pi$

91. $s = 12 \cdot \dfrac{\pi}{4} = 3\pi$ ft

93. $s = 4000 \cdot \dfrac{3\pi}{180} \approx 209.4$ miles

95. radius is $r = \dfrac{s}{\alpha} = \dfrac{1}{1} = 1$ mile.

97. radius is $r = \dfrac{s}{\alpha} = \dfrac{10}{\pi} \approx 3.18$ km

99. Let α be the central angle, in radian measure, determined by the towers of the bridge. Since the length of the arc joining the bases of the towers is 4260 ft, we obtain

$$\begin{aligned} s &= \alpha r \\ 4260 &= \alpha(4000 \cdot 5280) \\ \alpha &= \dfrac{4260}{4000(5280)}. \end{aligned}$$

Let s_t be the length of the arc joining the tops of the towers. Since the towers are 693 ft tall, we have

$$s_t = \alpha(4000 \cdot 5280 + 693).$$

Thus, the length of the arc joining the tops of the towers is greater than the length of the arc joining the bases by the amount shown below.

$$s_t - 4260 = 0.1397...\text{ft} \approx 1.68 \text{ in}$$

101. Since $7° \approx 0.12217305$, the radius of the earth according to Eratosthenes is

$$r = \dfrac{s}{\alpha} \approx \dfrac{800}{0.12217305} \approx 6548.089 \text{ km}.$$

Consequently, the circumference is

$$2\pi r \approx 41,143 \text{ km}.$$

However, if $r = 6378$ km, then the circumference is 40,074 km.

For Thought

1. False, since $\cos 90° = 0$.

2. False, since $\cos 90 \approx -0.4$.

3. True, since $\sin(45°) = \dfrac{\sqrt{2}}{2} = \dfrac{1}{\sqrt{2}}$.

4. False, since $\sin\left(-\dfrac{\pi}{3}\right) = -\dfrac{\sqrt{3}}{2}$ and

$\sin\left(\dfrac{\pi}{3}\right) = \dfrac{\sqrt{3}}{2}$.

5. False, since $\cos\left(-\dfrac{\pi}{3}\right) = \dfrac{1}{2}$ and

$-\cos\left(\dfrac{\pi}{3}\right) = -\dfrac{1}{2}$.

6. True, since the reference arc of $390°$ is $30°$ and $390°$ lies in Quadrant I.

7. False, since α lies in Quadrant IV.

8. False, since $\sin(\alpha) = -\dfrac{1}{2}$.

9. False

10. True, since $(1 - \sin(\alpha))(1 + \sin(\alpha)) = 1 - \sin^2(\alpha) = \cos^2(\alpha)$.

3.2 Exercises

1. $(1,0), \left(\dfrac{\sqrt{2}}{2}, \dfrac{\sqrt{2}}{2}\right), (0,1), \left(-\dfrac{\sqrt{2}}{2}, \dfrac{\sqrt{2}}{2}\right),$

$(-1,0), \left(-\dfrac{\sqrt{2}}{2}, -\dfrac{\sqrt{2}}{2}\right), (0,-1), \left(\dfrac{\sqrt{2}}{2}, -\dfrac{\sqrt{2}}{2}\right)$

3. 0 **5.** 0

7. 0 **9.** 0

11. $\dfrac{\sqrt{2}}{2}$ **13.** $-\dfrac{\sqrt{2}}{2}$

15. $\dfrac{1}{2}$ **17.** $\dfrac{1}{2}$

19. $-\dfrac{\sqrt{3}}{2}$ **21.** $\dfrac{\sqrt{3}}{2}$

23. $\sin(390°) = \sin(30°) = \dfrac{1}{2}$

25. $\cos(-420°) = \cos(300°) = \dfrac{1}{2}$

27. $\cos\left(\dfrac{13\pi}{6}\right) = \cos\left(\dfrac{\pi}{6}\right) = \dfrac{\sqrt{3}}{2}$

29. $\dfrac{\sqrt{2}}{2}$

31. -1

33. $\dfrac{\sqrt{3}}{2}$

35. $\dfrac{1}{2}$

37. $\dfrac{\cos(\pi/3)}{\sin(\pi/3)} = \dfrac{1/2}{\sqrt{3}/2} = \dfrac{1}{\sqrt{3}} = \dfrac{\sqrt{3}}{3}$

39. $\dfrac{\sin(7\pi/4)}{\cos(7\pi/4)} = \dfrac{-\sqrt{2}/2}{\sqrt{2}/2} = -1$

41. $\sin\left(\dfrac{\pi}{3} + \dfrac{\pi}{6}\right) = \sin\left(\dfrac{\pi}{2}\right) = 1$

43. $\dfrac{1 - \cos(5\pi/6)}{\sin(5\pi/6)} = \dfrac{1 - (-\sqrt{3}/2)}{1/2} \cdot \dfrac{2}{2} = 2 + \sqrt{3}$

$\dfrac{1}{2 - \sqrt{3}} \cdot \dfrac{2 + \sqrt{3}}{2 + \sqrt{3}} = \dfrac{2 + \sqrt{3}}{4 - 3} = 2 + \sqrt{3}$

45. $\dfrac{\sqrt{2}}{2} + \dfrac{\sqrt{2}}{2} = \sqrt{2}$

47. $+$, since sine is positive in the 2nd quadrant

49. $+$, since cosine is positive in the 4th quadrant

51. $-$, since sine is negative in the 3rd quadrant

53. $-$, since cosine is negative in the 3rd quadrant

55. 0.9999

57. 0.4035

59. -0.7438

61. 1.0000

63. -0.2588

65. $\sin\left(\dfrac{\pi}{2}\right) = 1$

67. $\cos\left(\dfrac{\pi}{3}\right) = \dfrac{1}{2}$

69. $\sin\left(\dfrac{3\pi}{4}\right) = \dfrac{\sqrt{2}}{2}$

71. $\cos\left(\dfrac{\pi}{6}\right) = \dfrac{\sqrt{3}}{2}$

73. Use the Fundamental Identity.

$$\left(\frac{5}{13}\right)^2 + \cos^2(\alpha) = 1$$

$$\frac{25}{169} + \cos^2(\alpha) = 1$$

$$\cos^2(\alpha) = \frac{144}{169}$$

$$\cos(\alpha) = \pm\frac{12}{13}$$

Since α is in quadrant II, $\cos(\alpha) = -12/13$.

75. Use the Fundamental Identity.

$$\left(\frac{3}{5}\right)^2 + \sin^2(\alpha) = 1$$

$$\frac{9}{25} + \sin^2(\alpha) = 1$$

$$\sin^2(\alpha) = \frac{16}{25}$$

$$\sin(\alpha) = \pm\frac{4}{5}$$

Since α is in quadrant IV, $\sin(\alpha) = -4/5$.

77. Use the Fundamental Identity.

$$\left(\frac{1}{3}\right)^2 + \cos^2(\alpha) = 1$$

$$\frac{1}{9} + \cos^2(\alpha) = 1$$

$$\cos^2(\alpha) = \frac{8}{9}$$

$$\cos(\alpha) = \pm\frac{2\sqrt{2}}{3}$$

Since $\cos(\alpha) > 0$, $\cos(\alpha) = \dfrac{2\sqrt{2}}{3}$.

79. Since $x(t) = 4\sin(t) - 3\cos(t)$,
the location of the weight after 3 seconds
is $x(3) = 4\sin(3) - 3\cos(3) \approx 3.53$ cm.
It is below its equilibrium position.

81. The angle between the tips of two adjacent

teeth is $\dfrac{2\pi}{22} = \dfrac{\pi}{11}$. The actual distance is

$c = 6\sqrt{2 - 2\cos(\pi/11)} \approx 1.708$ in.

The length of the arc is $s = 6 \cdot \dfrac{\pi}{11} \approx 1.714$ in.

For Thought

1. False, the period is 1.

2. False, the range is $[-1, 7]$.

3. False, the phase shift is $-\pi/12$.

4. True

5. True

6. True, $\dfrac{2\pi}{0.1\pi} = 20$.

7. False

8. True, since $\dfrac{2\pi}{b} = \dfrac{2\pi}{4} = \dfrac{\pi}{2}$.

9. True

10. True

3.3 Exercises

1. $y = -2\sin x$, amplitude 2

3. $y = 3\cos x$, amplitude 3

5. Amplitude 2, period 2π, phase shift 0

7. Amplitude 1, period 2π, phase shift $\pi/2$

9. Amplitude 2, period 2π, phase shift $-\pi/3$

11. Amplitude 1, phase shift 0, some points are

$$(0, 0), \left(\frac{\pi}{2}, -1\right), (\pi, 0), \left(\frac{3\pi}{2}, 1\right), (2\pi, 0)$$

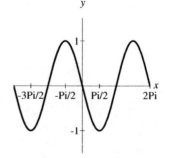

13. Amplitude 3, phase shift 0, some points are

$(0,0)$, $\left(\dfrac{\pi}{2}, -3\right)$, $(\pi, 0)$, $\left(\dfrac{3\pi}{2}, 3\right)$, $(2\pi, 0)$

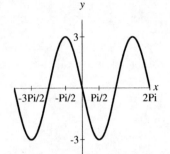

15. Amplitude 1/2, phase shift 0, some points are

$(0, 1/2)$, $(\pi/2, 0)$, $(\pi, -1/2)$ $(3\pi/2, 0)$,
$(2\pi, 1/2)$

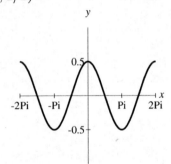

17. Amplitude 1, phase shift $-\pi$, some points are

$(0,0)$, $\left(\dfrac{\pi}{2}, -1\right)$, $(\pi, 0)$, $\left(\dfrac{3\pi}{2}, 1\right)$, $(2\pi, 0)$

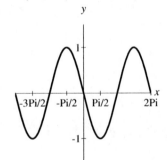

19. Amplitude 1, phase shift $\pi/3$, some points are

$\left(-\dfrac{2\pi}{3}, -1\right)$, $\left(-\dfrac{\pi}{6}, 0\right)$, $\left(\dfrac{\pi}{3}, 1\right)$, $\left(\dfrac{5\pi}{6}, 0\right)$,
$\left(\dfrac{4\pi}{3}, -1\right)$

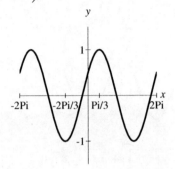

21. Amplitude 1, phase shift 0, some points are

$(0,3)$, $\left(\dfrac{\pi}{2}, 2\right)$, $(\pi, 1)$ $\left(\dfrac{3\pi}{2}, 2\right)$, $(2\pi, 3)$

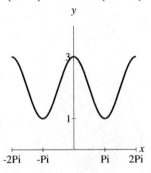

23. Amplitude 1, phase shift 0, some points are

$(0, -1)$, $\left(\dfrac{\pi}{2}, -2\right)$, $(\pi, -1)$, $\left(\dfrac{3\pi}{2}, 0\right)$, $(2\pi, -1)$

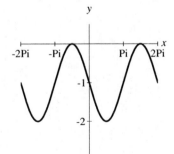

25. Amplitude 1, phase shift $-\pi/4$, some points

are $\left(-\dfrac{\pi}{4}, 2\right)$, $\left(\dfrac{\pi}{4}, 3\right)$, $\left(\dfrac{3\pi}{4}, 2\right)$, $\left(\dfrac{5\pi}{4}, 1\right)$,

$\left(\dfrac{7\pi}{4}, 2\right)$

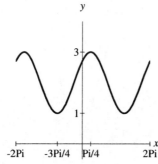

27. Amplitude 2, phase shift $-\pi/6$, some points

are $\left(-\dfrac{\pi}{6}, 3\right)$, $\left(\dfrac{\pi}{3}, 1\right)$, $\left(\dfrac{5\pi}{6}, -1\right)$, $\left(\dfrac{4\pi}{3}, 1\right)$,

$\left(\dfrac{11\pi}{6}, 3\right)$

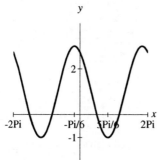

29. Amplitude 2, phase shift $\pi/3$, some points are

$\left(-\dfrac{\pi}{6}, 3\right)$, $\left(\dfrac{\pi}{3}, 1\right)$, $\left(\dfrac{5\pi}{6}, -1\right)$, $\left(\dfrac{4\pi}{3}, 1\right)$,

$\left(\dfrac{11\pi}{6}, 3\right)$

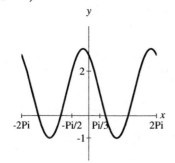

31. Amplitude 3, period $\pi/2$, phase shift 0

33. Amplitude 1, period $\dfrac{2\pi}{1/2}$ or 4π, phase shift 0

35. Amplitude 2, period 2π, phase shift π

37. Amplitude 2; since $y = -2\cos(2(x+\pi/4))$, the period is π and the phase shift is $-\pi/4$

39. Amplitude 2; since $y = -2\cos\left(\dfrac{\pi}{2}(x+2)\right)$,

the period is $\dfrac{2\pi}{\pi/2}$ or 4, and the phase shift

is -2

41. Note, $A = (7-3)/2 = 2$. Since $A + D = 2 + D = 7$, we find that $D = 5$. Since $C = -\pi/2$ and $2\pi/B = \pi$, we obtain $B = 2$. Thus,

$$y = 2\sin\left(2\left(x+\dfrac{\pi}{2}\right)\right) + 5.$$

43. Note, $A = (9-(-1))/2 = 5$. Since $A + D = 5 + D = 9$, we obtain that $D = 4$. Since $C = 2$ and $2\pi/B = 2$, we find $B = \pi$. Thus,

$$y = 5\sin\left(\pi\left(x-2\right)\right) + 4.$$

45. Note, $A = (3-(-9))/2 = 6$. Since $A + D = 6 + D = 3$, we find that $D = -3$. Since $C = -\pi$ and $2\pi/B = 1/2$, we find $B = 4\pi$. Hence,

$$y = 6\sin\left(4\pi\left(x+\pi\right)\right) - 3.$$

47. $y = -\sin\left(x - \dfrac{\pi}{4}\right) + 1$

49. $y = -\left[3\cos\left(x-\pi\right) - 2\right]$ or

equivalently $y = -3\cos\left(x-\pi\right) + 2$

51. Period $2\pi/3$, phase shift 0, range $[-1, 1]$,

labeled points are $(0, 0)$, $\left(\dfrac{\pi}{6}, 1\right)$,

$\left(\dfrac{\pi}{3}, 0\right)$, $\left(\dfrac{\pi}{2}, -1\right)$, $\left(\dfrac{2\pi}{3}, 0\right)$

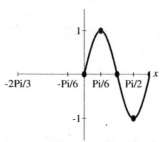

53. Period π, phase shift 0, range $[-1, 1]$, labeled points are $(0, 0)$, $\left(\dfrac{\pi}{4}, -1\right)$, $\left(\dfrac{\pi}{2}, 0\right)$, $\left(\dfrac{3\pi}{4}, 1\right)$, $(\pi, 0)$

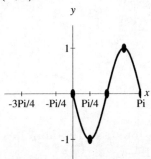

55. Period $\pi/2$, phase shift 0, range $[1, 3]$, labeled points are $(0, 3)$, $\left(\dfrac{\pi}{8}, 2\right)$, $\left(\dfrac{\pi}{4}, 1\right)$, $\left(\dfrac{3\pi}{8}, 2\right)$, $\left(\dfrac{\pi}{2}, 3\right)$

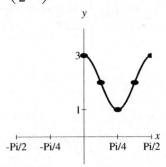

57. Period 8π, phase shift 0, range $[1, 3]$, labeled points are $(0, 2)$, $(2\pi, 1)$, $(4\pi, 2)$, $(6\pi, 3)$, $(8\pi, 2)$

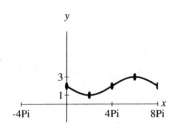

59. Period 6, phase shift 0, range $[-1, 1]$, labeled points are $(0, 0)$, $(1.5, 1)$, $(3, 0)$, $(4.5, -1)$, $(6, 0)$

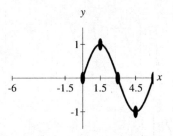

61. Period π, phase shift $\pi/2$, range $[-1, 1]$, labeled points are $\left(\dfrac{\pi}{2}, 0\right)$, $\left(\dfrac{3\pi}{4}, 1\right)$, $(\pi, 0)$, $\left(\dfrac{5\pi}{4}, -1\right)$, $\left(\dfrac{3\pi}{2}, 0\right)$

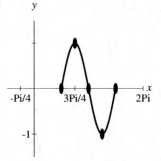

63. Period 4, phase shift -3, range $[-1, 1]$, labeled points are $(-3, 0)$, $(-2, 1)$, $(-1, 0)$, $(0, -1)$, $(1, 0)$

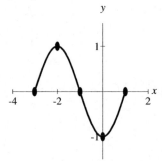

65. Period π, phase shift $-\pi/6$, range $[-1, 3]$,

labeled points are $\left(-\dfrac{\pi}{6}, 3\right)$, $\left(\dfrac{\pi}{12}, 1\right)$,

$\left(\dfrac{\pi}{3}, -1\right)$, $\left(\dfrac{7\pi}{12}, 1\right)$, $\left(\dfrac{5\pi}{6}, 3\right)$

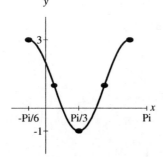

67. Period $\dfrac{2\pi}{3}$, phase shift $\dfrac{\pi}{6}$, range $\left[-\dfrac{3}{2}, -\dfrac{1}{2}\right]$,

labeled points are $\left(\dfrac{\pi}{6}, -1\right)$, $\left(\dfrac{\pi}{3}, -\dfrac{3}{2}\right)$,

$\left(\dfrac{\pi}{2}, -1\right)$, $\left(\dfrac{2\pi}{3}, -\dfrac{1}{2}\right)$, $\left(\dfrac{5\pi}{6}, -1\right)$

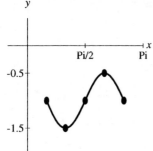

69. $y = 2\sin\left(2\left(x - \dfrac{\pi}{4}\right)\right)$

71. $y = 3\sin\left(\dfrac{3}{2}\left(x + \dfrac{\pi}{3}\right)\right) + 3$

73. 100 cycles per second

75. Frequency is $\dfrac{1}{0.025} = 40$ cycles per hour

77. Substitute $v_o = 6$, $\omega = 2$, and $x_o = 0$ into

$$x(t) = \dfrac{v_o}{\omega} \cdot \sin(\omega t) + x_o \cdot \cos(\omega t). \text{ Then}$$

$$x(t) = 3\sin(2t).$$

The amplitude is 3 and period is π.

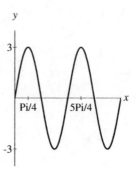

79. 11 years

81. Note that the maximum and minimum values of $\sin(t)$ are 1 and -1.
(a) Maximum volume is 1300 cc and miminum volume is 500 cc
(b) The runner takes a breath every 1/30 (which is the period) of a minute. So a runner makes 30 breaths in one minute.

83. Period is 12, amplitude is 15,000, phase-shift is -3, vertical translation is 25,000, a formula for the curve is

$$y = 15,000 \sin\left(\dfrac{\pi}{6}x + \dfrac{\pi}{2}\right) + 25,000.$$

For April (when $x = 4$), the revenue is $15,000 \sin\left(\dfrac{\pi}{6}x + \dfrac{\pi}{2}\right) + 25,000 \approx \$17,500$.

85. Since the period is 20, the amplitude is 1, and the vertical translation is 1, an equation for the swell is $y = \sin\left(\dfrac{\pi}{10}x\right) + 1$.

For Thought

1. True, since $\sin(\pi/4) = \cos(\pi/4)$.

2. False, since $\cot(\pi/2) = 0$ and $\dfrac{1}{\tan(\pi/2)}$ is undefined for $\tan(\pi/2)$ is undefined.

3. True, $\csc(60°) = \dfrac{2}{\sqrt{3}} \cdot \dfrac{\sqrt{3}}{\sqrt{3}} = \dfrac{2\sqrt{3}}{3}$.

4. False, since $\tan(5\pi/2)$ is undefined.

5. False, $\sec(95°) < 0$.

6. True, since $\sin 120° = \dfrac{\sqrt{3}}{2}$ and $\csc 120° = 1/\sin 120°$.

7. False, since the ranges are $(-\infty, -2] \cup [2, \infty)$ and $(-\infty, -0.5] \cup [0.5, \infty)$, respectively.

8. True, since $|\csc x| \geq 1$ and $|0.5 \csc x| \geq 0.5$.

9. True, since $\tan\left(3 \cdot \dfrac{\pm\pi}{6}\right) = \tan\left(\pm\dfrac{\pi}{2}\right)$ is undefined.

10. True, since $\cot\left(4 \cdot \dfrac{\pm\pi}{4}\right) = \cot(\pm\pi)$ is undefined.

3.4 Exercises

1. $\tan(0) = 0$, $\tan(\pi/4) = 1$, $\tan(\pi/2)$ undefined, $\tan(3\pi/4) = -1$, $\tan(\pi) = 0$, $\tan(5\pi/4) = 1$, $\tan(3\pi/2)$ undefined, $\tan(7\pi/4) = -1$

3. $\sqrt{3}$ **5.** -1

7. 0   **9.** $-\sqrt{3}/3$

11. $\dfrac{2\sqrt{3}}{3}$ **13.** Undefined

15. Undefined   **17.** $\sqrt{2}$

19. -1 **21.** $\sqrt{3}$

23. -2 **24.** undefined

25. $-\sqrt{2}$ **27.** 0

29. 48.0785   **31.** -2.8413

33. 500.0003   **35.** 1.0353   **37.** 636.6192

39. -1.4318 **41.** 71.6221   **43.** -0.9861

45. $\sec^2\left(2\left(\dfrac{\pi}{6}\right)\right) = \sec^2\left(\dfrac{\pi}{3}\right) = 2^2 = 4$

47. $\tan\left(\dfrac{\frac{\pi}{3}}{2}\right) = \tan\left(\dfrac{\pi}{6}\right) = \dfrac{\sqrt{3}}{3}$

49. $\sec\left(\dfrac{\frac{3\pi}{2}}{2}\right) = \sec\left(\dfrac{3\pi}{4}\right) = -\sqrt{2}$

51. $y = \tan(3x)$ has period $\pi/3$

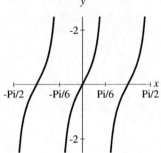

53. $y = \cot(x + \pi/4)$ has period π

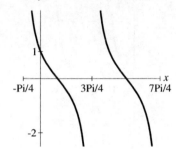

55. $y = \cot(x/2)$ has period 2π

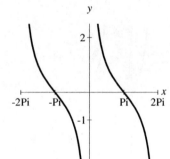

57. $y = \tan(\pi x)$ has period 1

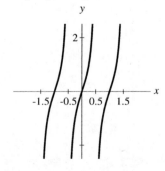

59. $y = -2\tan(x)$ has period π

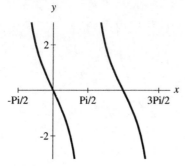

61. $y = -\cot(x + \pi/2)$ has period π

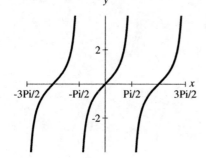

63. $y = \cot(2x - \pi/2)$ has period $\pi/2$

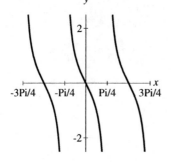

65. $y = \tan\left(\dfrac{\pi}{2} \cdot x - \dfrac{\pi}{2}\right)$ has period 2

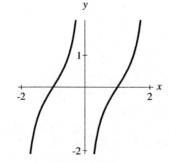

67. period π, range $(-\infty, -1] \cup [1, \infty)$

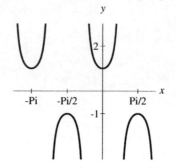

69. period 2π, $(-\infty, -1] \cup [1, \infty)$

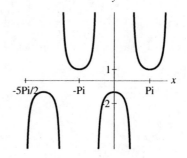

71. period 4π, $(-\infty, -1] \cup [1, \infty)$

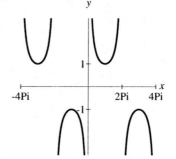

73. period 4, $(-\infty, -1] \cup [1, \infty)$

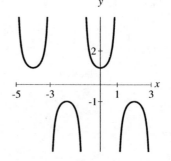

75. period 2π, $(-\infty, -2] \cup [2, \infty)$

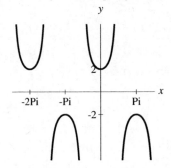

77. period π, $(-\infty, -1] \cup [1, \infty)$

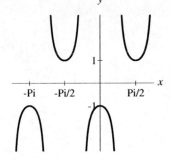

79. period 4, $(-\infty, -1] \cup [1, \infty)$

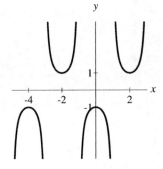

81. period π, $(-\infty, 0] \cup [4, \infty)$

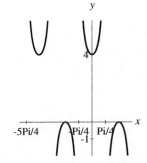

83. Period $\pi/B = \pi/2 = \pi$, range $(-\infty, \infty)$

85. Period $2\pi/B = 2\pi/(1/2) = 4\pi$,
range $(-\infty, -2 - 1] \cup [2 - 1, \infty)$ or
$(-\infty, -3] \cup [1, \infty)$

87. Period $2\pi/B = 2\pi/2 = \pi$,
range $(-\infty, -3 - 4] \cup [3 - 4, \infty)$ or
$(-\infty, -7] \cup [-1, \infty)$

89. $y = 3\tan\left(x - \dfrac{\pi}{4}\right) + 2$

91. $y = -\sec(x + \pi) + 2$

For Thought

1. True, $\sin^{-1}(0) = 0 = \sin(0)$.

2. True, since $\sin(3\pi/4) = \dfrac{\sqrt{2}}{2} = \dfrac{1}{\sqrt{2}}$.

3. False, $\cos^{-1}(0) = \pi/2$.

4. False, $\sin^{-1}(\sqrt{2}/2) = 45°$.

5. False, since it equals $\tan^{-1}(1/5)$.

6. True, since $1/5 = 0.2$.

7. True, $\sin(\cos^{-1}(\sqrt{2}/2)) = \sin(\pi/4) = 1/\sqrt{2}$.

8. True by definition of $y = \sec^{-1}(x)$.

9. False, since $f^{-1}(x) = \sin(x)$ where
$-\pi/2 \le x \le \pi/2$.

10. False, the secant and cosecant functions are not one-to-one functions and therefore do not have inverse functions.

3.5 Exercises

1. $-\pi/6$ **3.** $\pi/6$

5. $\pi/4$ **7.** $-45°$

9. $30°$ **11.** $0°$

13. $-19.5°$ **15.** $34.6°$

17. $3\pi/4$ **19.** $\pi/3$

21. π **23.** $135°$

25. $180°$ **27.** $120°$

29. $173.2°$　**31.**　$89.9°$

33. $-\pi/4$　**35.**　$\pi/3$

37. $\pi/4$　**39.**　$-\pi/6$

41. 0　**43.**　$\pi/2$

45. $3\pi/4$　**47.**　$2\pi/3$

49. 0.60　**51.**　3.02　**53.**　-0.14

55. 1.87　**57.**　1.15　**59.**　-0.36

61. 3.06　**63.**　0.06

65. $\tan(\pi/3) = \sqrt{3}$

67. $\sin^{-1}(-1/2) = -\pi/6$

69. $\cot^{-1}(\sqrt{3}) = \pi/6$

71. $\arcsin(\sqrt{2}/2) = \pi/4$

73. $\tan(\pi/4) = 1$

75. $\cos^{-1}(0) = \pi/2$

77. $\cos(2 \cdot \pi/4) = \cos(\pi/2) = 0$

79. $\sin^{-1}(2 \cdot 1/2) = \sin^{-1}(1) = \pi/2$

81. Solving for y, one finds

$$\begin{aligned} x &= \sin(2y) \\ \sin^{-1}(x) &= 2y \\ y &= \frac{\sin^{-1}(x)}{2}. \end{aligned}$$

Then

$$f^{-1}(x) = 0.5\sin^{-1}(x).$$

As x takes the values in $\left[-\dfrac{\pi}{4}, \dfrac{\pi}{4}\right]$, $2x$ takes all the values in $\left[-\dfrac{\pi}{2}, \dfrac{\pi}{2}\right]$, and $f(x) = \sin(2x)$ takes all the values in $[-1, 1]$. Thus, the range of f is $[-1, 1]$ which is the domain of f^{-1}.

83. Solving for y, one obtains

$$\begin{aligned} x &= 3 + \tan(\pi y) \\ x - 3 &= \tan(\pi y) \\ \tan^{-1}(x-3) &= \pi y \\ \frac{\tan^{-1}(x-3)}{\pi} &= y. \end{aligned}$$

Thus, we find

$$f^{-1}(x) = \frac{\tan^{-1}(x-3)}{\pi}.$$

As x takes the values in $\left(-\dfrac{1}{2}, \dfrac{1}{2}\right)$, πx takes all the values in $\left(-\dfrac{\pi}{2}, \dfrac{\pi}{2}\right)$, $\tan(\pi x)$ takes all the values in $(-\infty, \infty)$, and $f(x) = 3 + \tan(\pi x)$ will take all the values in $(-\infty, \infty)$. Thus, the range of f is $(-\infty, \infty)$, which is the domain of f^{-1}.

85. Solving for y, one obtains

$$\begin{aligned} x &= \sin^{-1}\left(\frac{y}{2}\right) + 3 \\ x - 3 &= \sin^{-1}\left(\frac{y}{2}\right) \\ \sin(x-3) &= \frac{y}{2} \\ y &= 2\sin(x-3). \end{aligned}$$

Thus, we obtain

$$f^{-1}(x) = 2\sin(x-3).$$

As x takes the values in $[-2, 2]$, $\dfrac{x}{2}$ takes all the values in $[-1, 1]$, $\sin^{-1}\left(\dfrac{x}{2}\right)$ takes all the values in $\left[-\dfrac{\pi}{2}, \dfrac{\pi}{2}\right]$, and $f(x) = \sin^{-1}\left(\dfrac{x}{2}\right) + 3$ will take all the values in $\left[-\dfrac{\pi}{2} + 3, \dfrac{\pi}{2} + 3\right]$. Thus, the range of f is $\left[3 - \dfrac{\pi}{2}, 3 + \dfrac{\pi}{2}\right]$, which is the domain of f^{-1}.

87. Consider the right triangle with hypotenuse 2400, altitude 2000, and the angle between the hypotenuse and the altitude is $\dfrac{\theta}{2}$. Since

$$\cos\left(\frac{\theta}{2}\right) = \frac{2000}{2400},$$

we obtain

$$\begin{aligned} \theta &= 2\cos^{-1}\left(\frac{2000}{2400}\right) \\ &\approx 67.1°. \end{aligned}$$

Thus, the airplane is within the range of the gun for $\theta \approx 67.1°$.

For Thought

1. False, since $\sin(\alpha) = -10/\sqrt{125}$.

2. True, since

$$r = \sqrt{(-1)^2 + 2^2} = \sqrt{5}$$

we get

$$\sec\alpha = \frac{r}{x} = \frac{\sqrt{5}}{-1} = -\sqrt{5}.$$

3. False, α may not lie in $[-\pi/2, \pi/2]$.

4. False, α may not lie in $[0, \pi]$.

5. True, since the side opposite α is the side adjacent to β.

6. False, $c = \sqrt{20}$.

7. False, $\tan\beta = 1/3$.

8. True, $\tan(55°) = 8/b$.

9. True, the smallest angle is $\cos^{-1}(4/5)$.

10. False, $\sin(90°) = 1 \neq$ hyp/adj.

3.6 Exercises

1. $\sin(\alpha) = 4/5, \cos(\alpha) = 3/5, \tan(\alpha) = 4/3,$
$\csc(\alpha) = 5/4, \sec(\alpha) = 5/3, \cot(\alpha) = 3/4$

3. $\sin(\alpha) = 3\sqrt{10}/10, \cos(\alpha) = -\sqrt{10}/10,$
$\tan(\alpha) = -3, \csc(\alpha) = \sqrt{10}/3,$
$\sec(\alpha) = -\sqrt{10}, \cot(\alpha) = -1/3$

5. $\sin(\alpha) = -\sqrt{3}/3, \cos(\alpha) = -\sqrt{6}/3,$
$\tan(\alpha) = \sqrt{2}/2, \csc(\alpha) = -\sqrt{3},$
$\sec(\alpha) = -\sqrt{6}/2, \cot(\alpha) = \sqrt{2}$

7. $\sin(\alpha) = -1/2, \cos(\alpha) = \sqrt{3}/2,$
$\tan(\alpha) = -\sqrt{3}/3, \csc(\alpha) = -2,$
$\sec(\alpha) = 2\sqrt{3}/3, \cot(\alpha) = -\sqrt{3}$

9. $\sin(\alpha) = \sqrt{5}/5, \cos(\alpha) = 2\sqrt{5}/5, \tan(\alpha) = 1/2,$
$\sin(\beta) = 2\sqrt{5}/5, \cos(\beta) = \sqrt{5}/5, \tan(\beta) = 2$

11. $\sin(\alpha) = 3\sqrt{34}/34, \cos(\alpha) = 5\sqrt{34}/34,$
$\tan(\alpha) = 3/5, \sin(\beta) = 5\sqrt{34}/34,$
$\cos(\beta) = 3\sqrt{34}/34, \tan(\beta) = 5/3$

13. $\sin(\alpha) = 4/5, \cos(\alpha) = 3/5, \tan(\alpha) = 4/3,$
$\sin(\beta) = 3/5, \cos(\beta) = 4/5, \tan(\beta) = 3/4$

15. $\tan^{-1}(9/1.5) \approx 80.5°$

17. $\tan^{-1}(\sqrt{3}) = 60°$

19. $\tan^{-1}(6.3/4) \approx 1.0$

21. $\tan^{-1}(1/\sqrt{5}) \approx 0.4$

23. Form the right triangle with $\alpha = 60°$, $c = 20$.

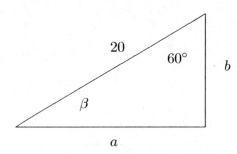

Since $\sin 60° = \dfrac{a}{20}$, we get

$$a = 20 \cdot \frac{\sqrt{3}}{2} = 10\sqrt{3}.$$

Since $\cos 60° = \dfrac{b}{20}$, we find

$$b = 20 \cdot \frac{1}{2} = 10.$$

Also, $\beta = 90° - 60° = 30°$.

25. Form the right triangle with $a = 6$, $b = 8$.

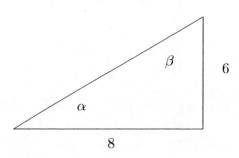

Note: $c = \sqrt{6^2 + 8^2} = 10$, $\tan(\alpha) = 6/8$,
so $\alpha = \tan^{-1}(6/8) \approx 36.9°$ and $\beta \approx 53.1°$.

27. Form the right triangle with $b = 6$, $c = 8.3$.

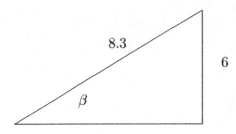

Note: $a = \sqrt{8.3^2 - 6^2} \approx 5.7$, $\sin(\beta) = 6/8.3$, so $\beta = \sin^{-1}(6/8.3) \approx 46.3°$ and $\alpha \approx 43.7°$.

29. Form the right triangle with $\alpha = 16°$, $c = 20$.

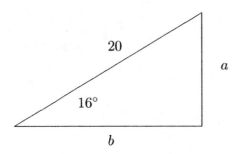

Since $\sin(16°) = a/20$ and $\cos(16°) = b/20$, $a = 20\sin(16°) \approx 5.5$ and $b = 20\cos(16°) \approx 19.2$. Also $\beta = 74°$.

31. Form the right triangle with $\alpha = 39°9'$, $a = 9$.

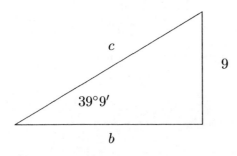

Since $\sin(39°9') = 9/c$ and $\tan(39°9') = 9/b$, then $c = 9/\sin(39°9') \approx 14.3$ and $b = 9/\tan(39°9') \approx 11.1$. Also $\beta = 50°51'$.

33. Let h be the height of the buliding.

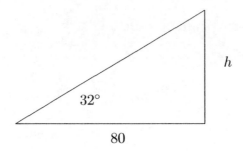

Since $\tan(32°) = h/80$, we obtain

$$h = 80 \cdot \tan(32°) \approx 50 \text{ ft.}$$

35. Let x be the distance between Muriel and the road at the time she encountered the swamp.

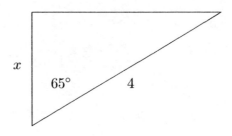

Since $\cos(65°) = x/4$, we find

$$x = 4 \cdot \cos(65°) \approx 1.7 \text{ miles.}$$

37. Let x be the distance between the car and a point on the highway directly below the observer.

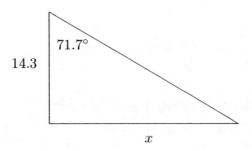

Since $\tan(71.7°) = x/14.3$, we obtain $x = 14.3 \cdot \tan(71.7°) \approx 43.2$ meters.

39. Let h be the height as in the picture below.

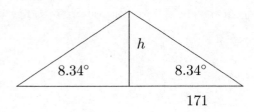

Since $\tan(8.34°) = h/171$, we obtain
$h = 171 \cdot \tan(8.34°) \approx 25.1$ ft.

41. Let α be the angle the guy wire makes with the ground.

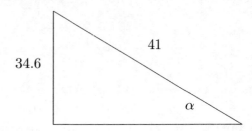

From the Pythagorean Theorem, the distance of the point to the base of the antenna is

$$\sqrt{41^2 - 34.6^2} \approx 22 \text{ meters.}$$

Also, $\alpha = \sin^{-1}(34.6/41) \approx 57.6°$.

43. Let h be the height.

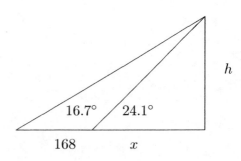

Note, $\tan 24.1° = \dfrac{h}{x}$ and $\tan 16.7° = \dfrac{h}{168 + x}$.

Solve for h in the second equation and substitute $x = \dfrac{h}{\tan 24.1°}$.

$$h = \tan(16.7°) \cdot \left(168 + \frac{h}{\tan 24.1°}\right)$$

$$h - \frac{h \tan(16.7°)}{\tan 24.1°} = \tan(16.7°) \cdot 168$$

$$h = \frac{168 \cdot \tan(16.7°)}{1 - \tan(16.7°)/\tan(24.1°)}$$

$$h \approx 153.1 \text{meters}$$

The height is 153.1 meters.

45. Let x be the closest distance the boat can come to the lighthouse LH.

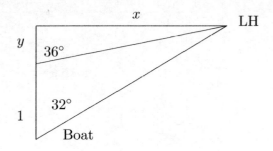

Since $\tan(36°) = x/y$ and $\tan(32°) = x/(1 + y)$, we obtain

46. Let h be the height of the Woolworth skyscraper.

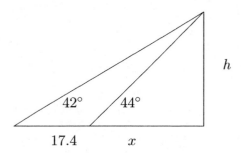

Since $\tan(44°) = h/x$ and

$$\tan(42°) = \frac{h}{17.4 + x}$$

we find

$$\tan(42°) = \frac{h}{17.4 + h/\tan(44°)}$$

$$17.4 \cdot \tan(42°) + \frac{\tan(42°)h}{\tan(44°)} = h$$

$$17.4 \cdot \tan(42°) = h\left(1 - \tan(42°)/\tan(44°)\right)$$

$$h = \frac{17.4 \cdot \tan(42°)}{1 - \tan(42°)/\tan(44°)}$$

$$h \approx 232 \text{ meters.}$$

47. In the triangle below CE stands for the center of the earth and PSE is a point on the surface of the earth on the horizon of the cameras of Landsat.

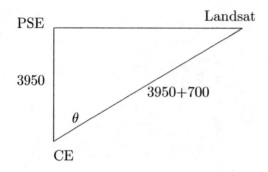

Since $\cos(\theta) = \dfrac{3950}{3950 + 700}$, we have

$$\theta = \cos^{-1}\left(\frac{3950}{3950 + 700}\right) \approx 0.5558 \text{ radians.}$$

But 2θ is the central angle, with vertex at CE, intercepted by the path on the surface of the earth as can be seen by Landsat. The width of this path is the arclength subtended by 2θ, that is,

$$s = r \cdot 2\theta = 3950 \cdot 2 \cdot 0.5558 \approx 4391 \text{ miles.}$$

49. Consider the right triangle formed by the hook, the center of the circle, and a point on the circle where the chain is tangent to the circle.

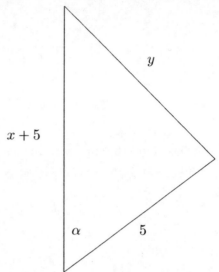

Then $\tan \alpha = \dfrac{y}{5}$ or $y = 5\tan \alpha$. Since the chain is 40 ft long and the angle $2\pi - 2\alpha$ intercepts an arc around the pipe where the chain wraps around the circle, we obtain $2y + 5(2\pi - 2\alpha) = 40$. By substitution, we get $10\tan \alpha + 10\pi - 10\alpha = 40$. With a graphing calculator, we obtain $\alpha \approx 1.09835$ radians. From the figure above, we get $\cos \alpha = \dfrac{5}{5 + x}$. Solving for x, we obtain $x = \dfrac{5 - 5\cos \alpha}{\cos \alpha} \approx 5.987$ ft.

For Thought

1. True, since

$$\tan x \cdot \cot x = \tan x \cdot \frac{1}{\tan x} = 1.$$

2. False, since

$$(\sin x + \cos x)^2 = \sin^2(x) + 2\sin(x)\cos(x) + \cos^2(x)$$
$$= 1 + 2\sin(x)\cos(x)$$

it follows that

$$(\sin x + \cos x)^2 \neq 1$$

and since

$$\sin^2 x + \cos^2 x = 1,$$

we obtain

$$(\sin x + \cos x)^2 \neq \sin^2 x + \cos^2 x.$$

3. False, since $\tan(1) = \sqrt{\sec^2(1) - 1}$.

4. False, since $\sin(-6) = -\sin(6)$ we get $\sin^2(-6) = (-\sin(6))^2 = \sin^2(6)$.

5. False, since $\dfrac{\pi}{4} - \dfrac{\pi}{3} = \dfrac{3\pi}{12} - \dfrac{4\pi}{12} = -\dfrac{\pi}{12}$.

6. True, by the sum identity for cosine.

7. True, since $\cos(\pi/2 - t) = \sin t$ for all t.

8. True, by an application of the sum identity for sine to $7\pi/12 = \pi/3 + \pi/4$.

9. False, $\sin\left(\dfrac{300°}{2}\right) = \sqrt{\dfrac{1 - \cos(300°)}{2}}$.

10. True, for if we apply a sum-to-product identity we obtain
$\cos(4) + \cos(12) =$
$$2\cos\left(\frac{4+12}{2}\right)\cos\left(\frac{4-12}{2}\right) =$$
$2\cos(8)\cos(-4) = 2\cos(8)\cos(4).$

3.7 Exercises

1. $\csc^2 x - \cot^2 x = 1$

3. $\dfrac{\sin^2 x(\sin^2 x - 1)}{1/\cos x} = \sin^2 x(-\cos^2 x) \cdot \cos x =$
$-\sin^2(x)\cos^3(x)$

5. $\dfrac{\cos w(\sin^2 w + \cos^2 w)}{\sec w} = \dfrac{\cos(w) \cdot 1}{\sec w} = \cos^2 w$

7. $\dfrac{\sin^2 x + \cos^2 x}{\sin x} = \dfrac{1}{\sin x} = \csc x$

9. $(-\sin x) \cdot (-\cot x) = \sin(x) \cdot \dfrac{\cos x}{\sin x} = \cos(x)$

11. $\dfrac{\sin(x)}{\cos(x)} + \dfrac{-\sin(x)}{\cos(x)} = 0$

13. $\cos(5)\cos(6) - \sin(5)\sin(6) = \cos(5+6) = \cos(11)$

15. $\cos(2k - k) = \cos(k)$

17. $\sin(23° + 67°) = \sin(90°) = 1$

19. $\sin(-\pi/2)\cos(\pi/5) + \cos(\pi/2)\sin(-\pi/5) = (-1)\cos(\pi/5) + (0)\sin(-\pi/5) = -\cos(\pi/5)$

21. $\tan\left(\dfrac{\pi}{9} + \dfrac{\pi}{6}\right) = \tan\left(\dfrac{5\pi}{18}\right)$

23. $\cos(3k)\cos(-k) - \sin(3k)\sin(-k) = \cos(3k + (-k)) = \cos(2k)$

25. $\sin(2 \cdot 13°) = \sin 26°$

27. $\dfrac{1}{2} \cdot \dfrac{2\tan 15°}{1 - \tan^2 15°} = \dfrac{1}{2} \cdot \tan(2 \cdot 15°) =$
$\dfrac{1}{2} \cdot \tan 30° = \dfrac{1}{2} \cdot \dfrac{\sqrt{3}}{3} = \dfrac{\sqrt{3}}{6}$

29. $2\sin\left(\dfrac{\pi}{9} - \dfrac{\pi}{2}\right)\cos\left(\dfrac{\pi}{9} - \dfrac{\pi}{2}\right) =$
$\sin\left(2 \cdot \left(\dfrac{\pi}{9} - \dfrac{\pi}{2}\right)\right) = \sin\left(\dfrac{2\pi}{9} - \pi\right) =$
$-\sin\left(\pi - \dfrac{2\pi}{9}\right) = -\sin(2\pi/9)$

31. Since $\sec\alpha = \sqrt{1 + (1/2)^2} = \sqrt{5}/2$, we get $\cos\alpha = 2/\sqrt{5} = 2\sqrt{5}/5$ and $\sin\alpha = \sqrt{1 - (2/\sqrt{5})^2} = 1/\sqrt{5} = \sqrt{5}/5$. Then $\csc\alpha = \sqrt{5}$ and $\cot\alpha = 2$.

33. Since $\sin\alpha = -\sqrt{1 - (-\sqrt{3}/5)^2} = -\sqrt{1 - 3/25} = -\sqrt{22}/5$, we obtain
$\csc\alpha = -5/\sqrt{22} = -5\sqrt{22}/22$,
$\sec\alpha = -5/\sqrt{3} = -5\sqrt{3}/3$,
$\tan\alpha = \dfrac{-\sqrt{22}/5}{-\sqrt{3}/5} = \sqrt{22}/\sqrt{3} = \sqrt{66}/3$,
and $\cot\alpha = \sqrt{3}/\sqrt{22} = \sqrt{66}/22$.

35. Since $\cos(2\alpha) = 2\cos^2\alpha - 1$, we get
$$\begin{aligned} 2\cos^2\alpha - 1 &= \frac{3}{5} \\ 2\cos^2\alpha &= \frac{8}{5} \\ \cos^2\alpha &= \frac{4}{5} \\ \cos\alpha &= \pm\frac{2}{\sqrt{5}}. \end{aligned}$$

But $0° < \alpha < 45°$, so $\cos \alpha = \dfrac{2}{\sqrt{5}} = \dfrac{2\sqrt{5}}{5}$

and $\sin \alpha = \sqrt{1 - \left(\dfrac{2}{\sqrt{5}}\right)^2} = \sqrt{1 - \dfrac{4}{5}} =$

$\sqrt{\dfrac{1}{5}} = \dfrac{1}{\sqrt{5}} = \dfrac{\sqrt{5}}{5}$.

Furthermore, $\sec \alpha = \dfrac{\sqrt{5}}{2}$, $\csc \alpha = \sqrt{5}$,

$\tan \alpha = \dfrac{1/\sqrt{5}}{2/\sqrt{5}} = \dfrac{1}{2}$, $\cot \alpha = 2$.

37. By a half-angle identity, we have

$$-\sqrt{\dfrac{1 + \cos \alpha}{2}} = -\dfrac{1}{4}$$
$$\dfrac{1 + \cos \alpha}{2} = \dfrac{1}{16}$$
$$1 + \cos \alpha = \dfrac{1}{8}$$
$$\cos \alpha = -\dfrac{7}{8}.$$

But $\pi \le \alpha \le 3\pi/2$,

so $\sin \alpha = -\sqrt{1 - \left(-\dfrac{7}{8}\right)^2} =$

$-\sqrt{1 - \dfrac{49}{64}} = -\sqrt{\dfrac{15}{64}} = -\dfrac{\sqrt{15}}{8}$.

Furthermore, $\sec \alpha = -\dfrac{8}{7}$, $\csc \alpha = -\dfrac{8\sqrt{15}}{15}$,

$\tan \alpha = \dfrac{-\sqrt{15}/8}{-7/8} = \dfrac{\sqrt{15}}{7}$, $\cot \alpha = \dfrac{7}{\sqrt{15}} =$

$\dfrac{7\sqrt{15}}{15}$.

39. $\sin\left(\dfrac{30°}{2}\right) = \sqrt{\dfrac{1 - \cos(30°)}{2}} =$

$\sqrt{\dfrac{1 - \sqrt{3}/2}{2} \cdot \dfrac{2}{2}} = \sqrt{\dfrac{2 - \sqrt{3}}{4}} = \dfrac{\sqrt{2 - \sqrt{3}}}{2}$

or by using a difference a formula one finds

$\sin(15°) = \sin(60° - 45°) = \dfrac{\sqrt{6} - \sqrt{2}}{4}$

41. $\cos(2\pi/3 - \pi/4) =$

$\cos(2\pi/3)\cos(\pi/4) + \sin(2\pi/3)\sin(\pi/4) =$

$-\dfrac{1}{2} \cdot \dfrac{\sqrt{2}}{2} + \dfrac{\sqrt{3}}{2} \cdot \dfrac{\sqrt{2}}{2} = \dfrac{\sqrt{6} - \sqrt{2}}{4}$

43. $\cos(-\pi/12) = \cos(\pi/12) = \cos(\pi/3 - \pi/4) =$
$\cos(\pi/3)\cos(\pi/4) + \sin(\pi/3)\sin(\pi/4) =$

$\dfrac{1}{2} \cdot \dfrac{\sqrt{2}}{2} + \dfrac{\sqrt{3}}{2} \cdot \dfrac{\sqrt{2}}{2} = \dfrac{\sqrt{2} + \sqrt{6}}{4}$

45. $\tan(45° + 30°) = \dfrac{\tan(45°) + \tan(30°)}{1 - \tan(45°)\tan(30°)} =$

$\dfrac{1 + \sqrt{3}/3}{1 - 1 \cdot \sqrt{3}/3} \cdot \dfrac{3}{3} = \dfrac{3 + \sqrt{3}}{3 - \sqrt{3}} \cdot \dfrac{3 + \sqrt{3}}{3 + \sqrt{3}} =$

$\dfrac{12 + 6\sqrt{3}}{9 - 3} = 2 + \sqrt{3}$

47. Rewrite the left side of the equation.

$$\tan(x)\cos(x) + \csc(x)\sin^2(x) =$$
$$\sin x + \sin x =$$
$$2\sin x$$

49.

$$2 - \csc(\beta)\sin(\beta) =$$
$$2 - 1 =$$
$$1 =$$
$$\sin^2(\beta) + \cos^2(\beta)$$

51.

$$\dfrac{\sec(x)}{\tan(x)} - \dfrac{\tan(x)}{\sec(x)} =$$
$$\dfrac{\sec^2(x) - \tan^2(x)}{\tan(x)\sec(x)} =$$
$$\dfrac{1}{\tan(x)\sec(x)} =$$
$$\cot(x)\cos(x)$$

53.

$$= \dfrac{\cos x + \csc x}{\cos x}$$
$$= \dfrac{\cos x}{\cos x} + \dfrac{\csc x}{\cos x}$$
$$1 + \csc x \sec x$$

55.

$$= \dfrac{1 + \sin(y)}{1 - \sin(y)} \cdot \dfrac{\csc(y)}{\csc(y)}$$
$$\dfrac{\csc(y) + 1}{\csc(y) - 1}$$

57.

$$= \frac{1 - (-\sin x)^2}{1 + \sin x}$$

$$= \frac{1 - \sin^2 x}{1 + \sin x}$$

$$= \frac{(1 - \sin x)(1 + \sin x)}{1 + \sin x}$$

$$1 - \sin(x)$$

59. We rewrite both sides:

$$\cos(x - \pi/2) = \cos x \cdot \frac{\sin x}{\cos x}$$

$$\cos(x)\cos(\pi/2) + \sin(x)\sin(\pi/2) =$$

$$\sin x = \sin x$$

61. In the proof, multiply each term by $\cos(\alpha - \beta)$. Also, the sum and difference identities for cosine expresses $\cos(\alpha + \beta)\cos(\alpha - \beta)$ as a difference of two squares.

$$\frac{\cos(\alpha + \beta)}{\cos \alpha + \sin \beta} =$$

$$\frac{\cos(\alpha + \beta)}{\cos \alpha + \sin \beta} \cdot \frac{\cos(\alpha - \beta)}{\cos(\alpha - \beta)} =$$

$$\frac{\cos^2 \alpha \cos^2 \beta - \sin^2 \alpha \sin^2 \beta}{(\cos \alpha + \sin \beta)\cos(\alpha - \beta)} =$$

$$\frac{\cos^2 \alpha(1 - \sin^2 \beta) - (1 - \cos^2 \alpha)\sin^2 \beta}{(\cos \alpha + \sin \beta)\cos(\alpha - \beta)} =$$

$$\frac{\cos^2 \alpha - \cos^2 \alpha \sin^2 \beta - \sin^2 \beta + \cos^2 \alpha \sin^2 \beta}{(\cos \alpha + \sin \beta)\cos(\alpha - \beta)} =$$

$$\frac{\cos^2 \alpha - \sin^2 \beta}{(\cos \alpha + \sin \beta)\cos(\alpha - \beta)} =$$

$$\frac{(\cos \alpha - \cos \beta)(\cos \alpha + \cos \beta)}{(\cos \alpha + \sin \beta)\cos(\alpha - \beta)} =$$

$$\frac{\cos \alpha - \sin \beta}{\cos(\alpha - \beta)} =$$

$$\frac{\cos \alpha - \sin \beta}{\cos(\beta - \alpha)}$$

63. We rewrite both sides:

$$\sin(180° - \alpha) = \frac{1 - \cos^2 \alpha}{\sin \alpha}$$

$$\sin(180°)\cos \alpha - \cos(180°)\sin \alpha =$$

$$\sin \alpha = \frac{\sin^2 \alpha}{\sin \alpha}$$

$$= \sin \alpha$$

65. In the proof, divide each term by $\cos \alpha \cos \beta$.

$$\frac{\cos(\alpha + \beta)}{\sin(\alpha - \beta)} =$$

$$\frac{\dfrac{\cos \alpha \cos \beta}{\cos \alpha \cos \beta} - \dfrac{\sin \alpha \sin \beta}{\cos \alpha \cos \beta}}{\dfrac{\sin \alpha \cos \beta}{\cos \alpha \cos \beta} - \dfrac{\cos \alpha \sin \beta}{\cos \alpha \cos \beta}} =$$

$$\frac{1 - \tan(\alpha)\tan(\beta)}{\tan(\alpha) - \tan(\beta)}$$

67.

$$\cos^4 s - \sin^4 s =$$

$$(\cos^2 s - \sin^2 s)(\cos^2 s + \sin^2 s) =$$

$$\cos(2s) \cdot (1) =$$

$$\cos(2s)$$

69.

$$\frac{\cos(2x) + \cos(2y)}{\sin(x) + \cos(y)} =$$

$$\frac{1 - 2\sin^2 x + 2\cos^2 y - 1}{\sin x + \cos y} =$$

$$2\frac{\cos^2 y - \sin^2 x}{\sin x + \cos y} =$$

$$2\frac{(\cos y - \sin x)(\cos y + \sin x)}{\sin x + \cos y} =$$

$$2\cos(y) - 2\sin(x)$$

71.

$$= \frac{\sin^2 u}{1 + \cos u}$$

$$= \frac{1 - \cos^2 u}{1 + \cos u}$$

$$= \frac{(1 - \cos u)(1 + \cos u)}{1 + \cos u}$$

$$= (1 - \cos u) \cdot \frac{2}{2}$$

$$= 2 \cdot \frac{1 - \cos u}{2}$$

$$2\sin^2(u/2)$$

73.

$$\frac{1 - \sin^2(x/2)}{1 + \sin^2(x/2)} =$$

$$\frac{1 - \left(\dfrac{1 - \cos x}{2}\right)}{1 + \left(\dfrac{1 - \cos x}{2}\right)} \cdot \frac{2}{2} =$$

$$\frac{2 - (1 - \cos x)}{2 + (1 - \cos x)} =$$

$$\frac{1 + \cos x}{3 - \cos x}$$

75. By using a sum-to-product identity, we find

$$\frac{\cos x - \cos(3x)}{\cos x + \cos(3x)} =$$

$$\frac{-2\sin\left(\dfrac{x + 3x}{2}\right)\sin\left(\dfrac{x - 3x}{2}\right)}{2\cos\left(\dfrac{x + 3x}{2}\right)\cos\left(\dfrac{x - 3x}{2}\right)} =$$

$$\frac{-2\sin(2x)\sin(-x)}{2\cos(2x)\cos(-x)} =$$

$$\frac{2\sin(2x)\sin x}{2\cos(2x)\cos x} =$$

$$\tan(2x)\tan(x)$$

77.

$$\frac{1}{2}\left[\cos(13° - 9°) - \cos(13° + 9°)\right] =$$

$$0.5\left[\cos 4° - \cos 22°\right]$$

79.

$$\frac{1}{2}\left[\cos\left(\frac{\pi}{6} - \frac{\pi}{5}\right) + \cos\left(\frac{\pi}{6} + \frac{\pi}{5}\right)\right] =$$

$$0.5\left[\cos\left(\frac{-\pi}{30}\right) + \cos\left(\frac{11\pi}{30}\right)\right] =$$

$$0.5\left[\cos\left(\frac{\pi}{30}\right) + \cos\left(\frac{11\pi}{30}\right)\right]$$

81.

$$\frac{1}{2}\left[\cos(52.5° - 7.5°) - \cos(52.5° + 7.5°)\right] =$$

$$\frac{1}{2}\left[\cos 45° - \cos 60°\right] =$$

$$\frac{1}{2}\left[\frac{\sqrt{2}}{2} - \frac{1}{2}\right] = \frac{\sqrt{2} - 1}{4}$$

83.

$$\frac{1}{2}\left[\sin\left(\frac{13\pi}{24} + \frac{5\pi}{24}\right) + \sin\left(\frac{13\pi}{24} - \frac{5\pi}{24}\right)\right] =$$

$$\frac{1}{2}\left[\sin(18\pi/24) + \sin(8\pi/24)\right] =$$

$$\frac{1}{2}\left[\sin(3\pi/4) + \sin(\pi/3)\right] =$$

$$\frac{1}{2}\left[\frac{\sqrt{2}}{2} + \frac{\sqrt{3}}{2}\right] = \frac{\sqrt{2} + \sqrt{3}}{4}$$

85.

$$2\cos\left(\frac{12° + 8°}{2}\right)\sin\left(\frac{12° - 8°}{2}\right) =$$

$$2\cos 10° \sin 2°$$

87.

$$-2\sin\left(\frac{\pi/3 + \pi/5}{2}\right)\sin\left(\frac{\pi/3 - \pi/5}{2}\right) =$$

$$-2\sin(4\pi/15)\sin(\pi/15) =$$

89.

$$2\sin\left(\frac{75° + 15°}{2}\right)\cos\left(\frac{75° - 15°}{2}\right) =$$

$$2\sin 45° \cos(30°) = 2 \cdot \frac{\sqrt{2}}{2}\frac{\sqrt{3}}{2} = \frac{\sqrt{6}}{2}$$

91.

$$-2\sin\left(\frac{\dfrac{-\pi}{24} + \dfrac{7\pi}{24}}{2}\right)\sin\left(\frac{\dfrac{-\pi}{24} - \dfrac{7\pi}{24}}{2}\right) =$$

$$-2\sin(3\pi/24)\sin(-4\pi/24) =$$
$$-2\sin(\pi/8)\sin(-\pi/6) =$$

$$-2\sin\left(\frac{\pi/4}{2}\right) \cdot \frac{-1}{2} = -2\sqrt{\frac{1 - \cos(\pi/4)}{2}} \cdot \frac{-1}{2}$$

$$= \sqrt{\frac{1 - \sqrt{2}/2}{2} \cdot \frac{2}{2}} = \sqrt{\frac{2 - \sqrt{2}}{4}} = \frac{\sqrt{2 - \sqrt{2}}}{2}$$

93. Since $a = 1$ and $b = -1$, we obtain
$r = \sqrt{1^2 + (-1)^2} = \sqrt{2}$. If the terminal side of α passes through $(1, -1)$, then
$\cos\alpha = a/r = 1/\sqrt{2}$ and
$\sin\alpha = b/r = -1/\sqrt{2}$. Choose $\alpha = -\pi/4$.

Thus, $\sin x - \cos x = r\sin(x + \alpha) = \sqrt{2}\sin(x - \pi/4)$.

95. Since $a = -1/2$ and $b = \sqrt{3}/2$, we obtain
$r = \sqrt{(-1/2)^2 + (\sqrt{3}/2)^2} = 1$. If the terminal side of α passes through

$(-1/2, \sqrt{3}/2)$, then $\cos\alpha = a/r = a/1 = a = -1/2$ and $\sin\alpha = b/r = b/1 = b = \sqrt{3}/2$. Choose $\alpha = 2\pi/3$. So $-\dfrac{1}{2}\sin x + \dfrac{\sqrt{3}}{2}\cos x = r\sin(x+\alpha) = \sin(x + 2\pi/3)$.

97. Since $a = \sqrt{3}/2$ and $b = -1/2$, we have $r = \sqrt{(\sqrt{3}/2)^2 + (-1/2)^2} = 1$. If the terminal side of α passes through $(\sqrt{3}/2, -1/2)$, then $\cos\alpha = a/r = a/1 = a = \sqrt{3}/2$ and $\sin\alpha = b/r = b/1 = b = -1/2$. Choose $\alpha = -\pi/6$. Thus,

$\dfrac{\sqrt{3}}{2}\sin x - \dfrac{1}{2}\cos x = r\sin(x+\alpha) = \sin(x - \pi/6)$.

99. Since the base of the TV screen is $b = d\cos\alpha$ and its height is $h = d\sin\alpha$, then the area A is given by

$$\begin{aligned}
A &= bh \\
&= (d\cos\alpha)(d\sin\alpha) \\
&= d^2\cos\alpha\sin\alpha \\
A &= \dfrac{d^2}{2}\sin(2\alpha).
\end{aligned}$$

101. Note that x can be written in the form $x = a\sin(t+\alpha)$. The maximum displacement of $x = \sqrt{3}\sin t + \cos t$ is $a = \sqrt{\sqrt{3}^2 + 1^2} = 2$. Thus, 2 meters is the maximum distance between the block and its resting position.

Since the terminal side of α goes through $(\sqrt{3}, 1)$, we get $\tan\alpha = 1/\sqrt{3}$ and one can choose $\alpha = \pi/6$. Then $x = 2\sin(t + \pi/6)$.

For Thought

1. False, the only solutions are $45°$ and $315°$.

2. False, there is no solution in $[0,\pi)$.

3. True, since $-29°$ and $331°$ are coterminal angles.

4. True   **5.**   True, since the right-side is a factorization of the left-side.

6. False, $x = 0$ is a solution to the first equation and not to the second equation.

7. False, $\cos^{-1}2$ is undefined.   **8.**   True

9. False, $x = 3\pi/4$ is not a solution to the first equation but is a solution to the second equation.

10 . False, rather $\left\{x | 3x = \dfrac{\pi}{2} + 2k\pi\right\} = \left\{x | x = \dfrac{\pi}{6} + \dfrac{2k\pi}{3}\right\}$.

3.8 Exercises

1. $\{x \mid x = \pi + 2k\pi, k \text{ an integer}\}$

3. $\{x \mid x = k\pi, k \text{ an integer}\}$

5. $\left\{x \mid x = \dfrac{3\pi}{2} + 2k\pi, k \text{ an integer}\right\}$

7. Solutions in $[0, 2\pi)$ are $x = \dfrac{\pi}{3}, \dfrac{5\pi}{3}$. So solution set is $\left\{x \mid x = \dfrac{\pi}{3} + 2k\pi \text{ or } x = \dfrac{5\pi}{3} + 2k\pi\right\}$.

9. Solutions in $[0, 2\pi)$ are $x = \dfrac{\pi}{4}, \dfrac{3\pi}{4}$. So solution set is $\left\{x \mid x = \dfrac{\pi}{4} + 2k\pi \text{ or } x = \dfrac{3\pi}{4} + 2k\pi\right\}$.

11. Solution in $[0, \pi)$ is $x = \dfrac{\pi}{4}$. The solution set is $\left\{x \mid x = \dfrac{\pi}{4} + k\pi\right\}$.

13. Solutions in $[0, 2\pi)$ are $x = \dfrac{5\pi}{6}, \dfrac{7\pi}{6}$. Then the solution set is $\left\{x \mid x = \dfrac{5\pi}{6} + 2k\pi \text{ or } x = \dfrac{7\pi}{6} + 2k\pi\right\}$.

15. Solutions in $[0, 2\pi)$ are $x = \dfrac{5\pi}{4}, \dfrac{7\pi}{4}$. The solution set is $\left\{x \mid x = \dfrac{5\pi}{4} + 2k\pi \text{ or } x = \dfrac{7\pi}{4} + 2k\pi\right\}$.

17. Solution in $[0, \pi)$ is $x = \dfrac{3\pi}{4}$. The solution set is $\left\{x \mid x = \dfrac{3\pi}{4} + k\pi\right\}$.

19. Solutions in $[0, 360°)$ are $\alpha = 90°, 270°$. So solution set is $\{\alpha \mid \alpha = 90° + k\cdot 180°\}$.

21. Solution in $[0, 360°)$ is $\alpha = 90°$. So the solution set is $\{\alpha \mid \alpha = 90° + k \cdot 360°\}$.

23. Solution in $[0, 180°)$ is $\alpha = 0°$.
The solution set is $\{\alpha \mid \alpha = k \cdot 180°\}$.

25. One solution is $\cos^{-1}(0.873) \approx 29.2°$. Another solution is $360° - 29.2° = 330.8°$. Solution set is $\{\alpha \mid \alpha = 29.2° + k360° \text{ or } \alpha = 330.8° + k360°\}$.

27. One solution is $\sin^{-1}(-0.244) \approx -14.1°$. This is coterminal with $345.9°$. Another solution is $180° + 14.1° = 194.1°$. Solution set is $\{\alpha \mid \alpha = 345.9° + k360° \text{ or } \alpha = 194.1° + k360°\}$.

29. One solution is $\tan^{-1}(5.42) \approx 79.5°$. Solution set is $\{\alpha \mid \alpha = 79.5° + k \cdot 180°\}$.

31. Values of $x/2$ in $[0, 2\pi)$ are $\pi/3$ and $5\pi/3$. Then we get

$$\frac{x}{2} = \frac{\pi}{3} + 2k\pi \text{ or } \frac{x}{2} = \frac{5\pi}{3} + 2k\pi$$

$$x = \frac{2\pi}{3} + 4k\pi \text{ or } x = \frac{10\pi}{3} + 4k\pi.$$

The solution set is

$$\left\{x \mid x = \frac{2\pi}{3} + 4k\pi \text{ or } x = \frac{10\pi}{3} + 4k\pi\right\}.$$

33. Value of $3x$ in $[0, 2\pi)$ is 0. Thus, $3x = 2k\pi$.
The solution set is $\left\{x \mid x = \frac{2k\pi}{3}\right\}$.

35. Since $\sin(x/2) = 1/2$, values of $x/2$ in $[0, 2\pi)$ are $\pi/6$ and $5\pi/6$. Then

$$\frac{x}{2} = \frac{\pi}{6} + 2k\pi \text{ or } \frac{x}{2} = \frac{5\pi}{6} + 2k\pi$$

$$x = \frac{\pi}{3} + 4k\pi \text{ or } x = \frac{5\pi}{3} + 4k\pi.$$

The solution set is

$$\left\{x \mid x = \frac{\pi}{3} + 4k\pi \text{ or } x = \frac{5\pi}{3} + 4k\pi\right\}.$$

37. Since $\sin(2x) = -\sqrt{2}/2$, values of $2x$ in $[0, 2\pi)$ are $5\pi/4$ and $7\pi/4$. Thus,

$$2x = \frac{5\pi}{4} + 2k\pi \text{ or } 2x = \frac{7\pi}{4} + 2k\pi$$

$$x = \frac{5\pi}{8} + k\pi \text{ or } x = \frac{7\pi}{8} + k\pi.$$

The solution set is

$$\left\{x \mid x = \frac{5\pi}{8} + k\pi \text{ or } x = \frac{7\pi}{8} + k\pi\right\}.$$

39. Value of $2x$ in $[0, \pi)$ is $\pi/3$. Then

$$2x = \frac{\pi}{3} + k\pi.$$

The solution set is $\left\{x \mid x = \frac{\pi}{6} + \frac{k\pi}{2}\right\}$.

41. Value of $4x$ in $[0, \pi)$ is 0. Then

$$4x = k\pi.$$

The solution set is $\left\{x \mid x = \frac{k\pi}{4}\right\}$.

43. The values of πx in $[0, 2\pi)$ are $\pi/6$ and $5\pi/6$. Then

$$\pi x = \frac{\pi}{6} + 2k\pi \text{ or } \pi x = \frac{5\pi}{6} + 2k\pi$$

$$x = \frac{1}{6} + 2k \text{ or } x = \frac{5}{6} + 2k.$$

The solution set is

$$\left\{x \mid x = \frac{1}{6} + 2k \text{ or } x = \frac{5}{6} + 2k\right\}.$$

45. Values of $2\pi x$ in $[0, 2\pi)$ are $\pi/2$ and $3\pi/2$. So

$$2\pi x = \frac{\pi}{2} + 2k\pi \text{ or } 2\pi x = \frac{3\pi}{2} + 2k\pi$$

$$x = \frac{1}{4} + k \text{ or } x = \frac{3}{4} + k.$$

The solution set is

$$\left\{x \mid x = \frac{1}{4} + \frac{k}{2}\right\}.$$

47. Since $\sin \alpha = -\sqrt{3}/2$, the solution set is $\{240°, 300°\}$.

49. Since $\cos 2\alpha = 1/\sqrt{2}$, values of 2α in $[0, 360°)$ are $45°$ and $315°$. Thus,

$$2\alpha = 45° + k \cdot 360° \text{ or } 2\alpha = 315° + k \cdot 360°$$

$$\alpha = 22.5° + k \cdot 180° \text{ or } \alpha = 157.5° + k \cdot 180°.$$

Then let $k = 0, 1$. The solution set is

$$\{22.5°, 157.5°, 202.5°, 337.5°\}.$$

51. Values of 3α in $[0, 360°)$ are $135°$ and $225°$. Then

$$3\alpha = 135° + k \cdot 360° \text{ or } 3\alpha = 225° + k \cdot 360°$$

$$\alpha = 45° + k \cdot 120° \text{ or } \alpha = 75° + k \cdot 120°.$$

By choosing $k = 0, 1, 2$, one obtains the solution set $\{45°, 75°, 165°, 195°, 285°, 315°\}$.

53. The value of $\alpha/2$ in $[0, 180°)$ is $30°$. Then

$$\frac{\alpha}{2} = 30° + k \cdot 180°$$

$$\alpha = 60° + k \cdot 360°.$$

By choosing $k = 0$, the solution set is $\{60°\}$.

55. A solution is $3\alpha = \sin^{-1}(0.34) \approx 19.88°$. Another solution is $3\alpha = 180° - 19.88° = 160.12°$. Then

$$3\alpha = 19.88° + k \cdot 360° \text{ or } 3\alpha = 160.12° + k \cdot 360°$$

$$\alpha \approx 6.6° + k \cdot 120° \text{ or } \alpha \approx 53.4° + k \cdot 120°.$$

Solution set is

$$\{\alpha \mid \alpha = 6.6° + k \cdot 120° \text{ or } \alpha = 53.4° + k \cdot 120°\}.$$

57. A solution is $3\alpha = \sin^{-1}(-0.6) \approx -36.87°$. This is coterminal with $323.13°$. Another solution is $3\alpha = 180° + 36.87° = 216.87°$. Then

$$3\alpha = 323.13° + k \cdot 360° \text{ or } 3\alpha = 216.87° + k \cdot 360°$$

$$\alpha \approx 107.7° + k \cdot 120° \text{ or } \alpha \approx 72.3° + k \cdot 120°.$$

The solution set is

$$\{\alpha \mid \alpha = 107.7° + k120° \text{ or } \alpha = 72.3° + k120°\}.$$

59. A solution is $2\alpha = \cos^{-1}(1/4.5) \approx 77.16°$. Another solution is $2\alpha = 360° - 77.16° = 282.84°$. Thus,

$$2\alpha = 77.16° + k \cdot 360° \text{ or } 2\alpha = 282.84° + k \cdot 360°$$

$$\alpha \approx 38.6° + k \cdot 180° \text{ or } \alpha \approx 141.4° + k \cdot 180°.$$

The solution set is

$$\{\alpha \mid \alpha = 38.6° + k180° \text{ or } \alpha = 141.4° + k180°\}.$$

61. A solution is $\alpha/2 = \sin^{-1}(-1/2.3) \approx -25.77°$. This is coterminal with $334.23°$. Another solution is $\alpha/2 = 180° + 25.77° = 205.77°$. Thus,

$$\frac{\alpha}{2} = 334.23° + k \cdot 360° \text{ or } \frac{\alpha}{2} = 205.77° + k \cdot 360°$$

$$\alpha \approx 668.5° + k \cdot 720° \text{ or } \alpha \approx 411.5° + k \cdot 720°.$$

The solution set is

$$\{\alpha \mid \alpha = 668.5° + k720° \text{ or } \alpha = 411.5° + k720°\}.$$

63. Set the right-hand side to zero and factor.

$$3\sin^2 x - \sin x = 0$$
$$\sin x(3\sin x - 1) = 0$$

Set each factor to zero.

$$\sin x = 0 \quad \text{or} \quad \sin x = 1/3$$
$$x = 0, \pi \quad \text{or} \quad x = \sin^{-1}(1/3) \approx 0.3$$

Another solution to $\sin x = 1/3$ is $x = \pi - 0.3 \approx 2.8$.
The solution set is $\{0, 0.3, 2.8, \pi\}$.

65. Set the right-hand side to zero and factor.

$$2\cos^2 x + 3\cos x + 1 = 0$$
$$(2\cos x + 1)(\cos x + 1) = 0$$

Set the factors to zero.

$$\cos x = -1/2 \quad \text{or} \quad \cos x = -1$$
$$x = 2\pi/3, 4\pi/3 \quad \text{or} \quad x = \pi$$

The solution set is $\{\pi, 2\pi/3, 4\pi/3\}$.

67. Substitute $\cos^2 x = 1 - \sin^2 x$.

$$5\sin^2 x - 2\sin x = 1 - \sin^2 x$$
$$6\sin^2 x - 2\sin x - 1 = 0$$

Apply the quadratic formula.

$$\sin x = \frac{2 \pm \sqrt{28}}{12}$$

$$\sin x = \frac{1 \pm \sqrt{7}}{6}$$

Then

$$x = \sin^{-1}\left(\frac{1+\sqrt{7}}{6}\right) \quad \text{or} \quad x = \sin^{-1}\left(\frac{1-\sqrt{7}}{6}\right)$$

$$x \approx 0.653 \quad \text{or} \quad x \approx -0.278.$$

Another solution is $\pi - 0.653 \approx 2.5$. An angle coterminal with -0.278 is $2\pi - 0.278 \approx 6.0$. Another solution is $\pi + 0.278 \approx 3.4$. The solution set is $\{0.7, 2.5, 3.4, 6.0\}$.

69. Squaring both sides of the equation, we obtain

$$\begin{aligned} \tan^2 x &= \sec^2 x - 2\sqrt{3}\sec x + 3 \\ \sec^2 x - 1 &= \sec^2 x - 2\sqrt{3}\sec x + 3 \\ -4 &= -2\sqrt{3}\sec x \\ \sec x &= 2/\sqrt{3} \\ x &= \pi/6, 11\pi/6. \end{aligned}$$

Checking $x = \pi/6$, one gets $\tan(\pi/6) = 1/\sqrt{3}$ and $\sec(\pi/6) - \sqrt{3} = 2/\sqrt{3} - \sqrt{3} = -1/\sqrt{3}$. Then $x = \pi/6$ is an extraneous root and the solution set is $\{11\pi/6\}$.

71. Square both sides of the equation.

$$\begin{aligned} \sin^2 x + 2\sqrt{3}\sin x + 3 &= 27\cos^2 x \\ \sin^2 x + 2\sqrt{3}\sin x + 3 &= 27(1 - \sin^2 x) \\ 28\sin^2 x + 2\sqrt{3}\sin x - 24 &= 0 \\ 14\sin^2 x + \sqrt{3}\sin x - 12 &= 0 \end{aligned}$$

By the quadratic formula, we get

$$\sin x = \frac{-\sqrt{3} \pm \sqrt{675}}{28}$$

$$\sin x = \frac{-\sqrt{3} \pm 15\sqrt{3}}{28}$$

$$\sin x = \frac{\sqrt{3}}{2}, \frac{-4\sqrt{3}}{7}.$$

Thus,

$$x = \frac{\pi}{3}, \frac{2\pi}{3} \quad \text{or} \quad x = \sin^{-1}\left(\frac{-4\sqrt{3}}{7}\right)$$

$$x = \frac{\pi}{3}, \frac{2\pi}{3} \quad \text{or} \quad x \approx -1.427.$$

Checking $x = 2\pi/3$, one finds $\sin(2\pi/3) + \sqrt{3} = \sqrt{3}/2 + \sqrt{3}$ and

$3\sqrt{3}\cos(2\pi/3)$ is a negative number. Then $x = 2\pi/3$ is an extraneous root.

An angle coterminal with -1.427 is $2\pi - 1.427 \approx 4.9$. In a similar way, one checks that $\pi + 1.427 \approx 4.568$ is an extraneous root. Since $\pi/3 \approx 1.0$, the solution set is $\{1.0, 4.9\}$.

73. Express the equation in terms of $\sin x$ and $\cos x$.

$$\begin{aligned} \frac{\sin x}{\cos x} \cdot 2\sin x \cos x &= 0 \\ 2\sin^2 x &= 0 \\ \sin x &= 0 \end{aligned}$$

Solution set is $\{0, \pi\}$.

75. Substitute the double-angle identity for $\sin x$.

$$\begin{aligned} 2\sin x \cos x - \sin x \cos x &= \cos x \\ \sin x \cos x - \cos x &= 0 \\ \cos x(\sin x - 1) &= 0 \\ \cos x = 0 \quad \text{or} \quad \sin x &= 1 \\ x = \pi/2, 3\pi/2 \quad \text{or} \quad x &= \pi/2 \end{aligned}$$

Solution set is $\{\pi/2, 3\pi/2\}$.

77. Use the sum identity for sine.

$$\begin{aligned} \sin(x + \pi/4) &= 1/2 \\ x + \frac{\pi}{4} = \frac{\pi}{6} + 2k\pi \quad \text{or} \quad x + \frac{\pi}{4} &= \frac{5\pi}{6} + 2k\pi \\ x = \frac{-\pi}{12} + 2k\pi \quad \text{or} \quad x &= \frac{7\pi}{12} + 2k\pi \end{aligned}$$

By choosing $k = 1$ in the first case and $k = 0$ in the second case, one finds the solution set is $\{23\pi/12, 7\pi/12\}$.

79. Apply the difference identity for sine.

$$\begin{aligned} \sin(2x - x) &= -1/2 \\ \sin x &= -1/2 \end{aligned}$$

The solution set is $\{7\pi/6, 11\pi/6\}$.

81. Use a half-angle identity for cosine and express equation in terms of $\cos\theta$.

$$\frac{1 + \cos\theta}{2} = \frac{1}{\cos\theta}$$

$$\cos\theta + \cos^2\theta = 2$$
$$\cos^2\theta + \cos\theta - 2 = 0$$
$$(\cos\theta + 2)(\cos\theta - 1) = 0$$
$$\cos\theta = -2 \quad \text{or} \quad \cos\theta = 1$$
$$\text{no solution} \quad \text{or} \quad \theta = 0°$$

Solution set is $\{0°\}$.

83. Dividing the equation by $2\cos\theta$, we get

$$\frac{\sin\theta}{\cos\theta} = \frac{1}{2}$$
$$\tan\theta = 0.5$$
$$\theta = \tan^{-1}(0.5) \approx 26.6°.$$

Another solution is $180° + 26.6° = 206.6°$.
Solution set is $\{26.6°, 206.6°\}$.

85. Express equation in terms of $\sin 3\theta$.

$$\sin 3\theta = \frac{1}{\sin 3\theta}$$
$$\sin^2 3\theta = 1$$
$$\sin 3\theta = \pm 1$$

Then

$$3\theta = 90° + k \cdot 360° \quad \text{or} \quad 3\theta = 270° + k \cdot 360°$$
$$\theta = 30° + k \cdot 120° \quad \text{or} \quad \theta = 90° + k \cdot 120°.$$

By choosing $k = 0, 1, 2$, one finds that the solution set is
$\{30°, 90°, 150°, 210°, 270°, 330°\}$.

87. By the method of completing the square, we get

$$\tan^2\theta - 2\tan\theta = 1$$
$$\tan^2\theta - 2\tan\theta + 1 = 2$$
$$(\tan\theta - 1)^2 = 2$$
$$\tan\theta - 1 = \pm\sqrt{2}$$
$$\theta = \tan^{-1}(1 + \sqrt{2}) \quad \text{or} \quad \theta = \tan^{-1}(1 - \sqrt{2})$$
$$\theta \approx 67.5° \quad \text{or} \quad \theta = -22.5°.$$

Other solutions are $180° + 67.5° = 247.5°$,
$180° - 22.5° = 157.5°$, and
$180° + 157.5° = 337.5°$. The solution set
is $\{67.5°, 157.5°, 247.5°, 337.5°\}$.

89. Factor as a perfect square.

$$(3\sin\theta + 2)^2 = 0$$
$$\sin\theta = -2/3$$
$$\theta = \sin^{-1}(-2/3) \approx -41.8°$$

An angle coterminal with $-41.8°$ is
$360° - 41.8° = 318.2°$. Another solution
is $180° + 41.8° = 221.8°$. The solution set
is $\{221.8°, 318.2°\}$.

91. By using the sum identity for tangent, we get

$$\tan(3\theta - \theta) = \sqrt{3}$$
$$2\theta = 60° + k \cdot 180°$$
$$\theta = 30° + k \cdot 90°.$$

By choosing $k = 1, 3$, one obtains that the
solution set is $\{120°, 300°\}$. Note, $30°$ and
$210°$ are not solutions.

93. Factoring, we get

$$(4\cos^2\theta - 3)(2\cos^2\theta - 1) = 0.$$

Then

$$\cos^2\theta = 3/4 \quad \text{or} \quad \cos^2\theta = 1/2$$
$$\cos\theta = \pm\sqrt{3}/2 \quad \text{or} \quad \cos\theta = \pm 1/\sqrt{2}.$$

The solution set is

$$\{30°, 45°, 135°, 150°, 210°, 225°, 315°, 330°\}.$$

95. Factoring, we obtain

$$(\sec^2\theta - 1)(\sec^2\theta - 4) = 0$$
$$\sec^2\theta = 1 \quad \text{or} \quad \sec^2\theta = 4$$
$$\sec\theta = \pm 1 \quad \text{or} \quad \sec\theta = \pm 2.$$

Solution set is $\{0°, 60°, 120°, 180°, 240°, 300°\}$.

97. Since $a = \sqrt{3}$ and $b = 1$, we obtain
$r = \sqrt{\sqrt{3}^2 + 1^2} = 2$. If the terminal side of
α goes through $(\sqrt{3}, 1)$, then $\tan\alpha = 1/\sqrt{3}$.
Then one can choose $\alpha = \pi/6$ and
$x = 2\sin(2t + \pi/6)$. The times when $x = 0$

are given by

$$\sin\left(2t + \frac{\pi}{6}\right) = 0$$

$$2t + \frac{\pi}{6} = k \cdot \pi$$

$$2t = -\frac{\pi}{6} + k \cdot \pi$$

$$t = -\frac{\pi}{12} + \frac{k \cdot \pi}{2}$$

$$t = -\frac{\pi}{12} + \frac{\pi}{2} + \frac{k \cdot \pi}{2}$$

$$t = \frac{5\pi}{12} + \frac{k \cdot \pi}{2}$$

where k is a nonnegative integer.

99. Since $v_o = 325$ and $d = 3300$, we have

$$325^2 \sin 2\theta = 32(3300)$$

$$\sin 2\theta = \frac{32(3300)}{325^2}$$

$$\sin 2\theta \approx 0.99976$$

$$2\theta \approx \sin^{-1}(0.99976)$$

$$2\theta \approx 88.74°$$

$$\theta \approx 44.4°.$$

Another angle is given by $2\theta = 180° - 88.74°$ $= 91.26°$ or $\theta = 91.26°/2 \approx 45.6°$.
The muzzle was aimed at $44.4°$ or $45.6°$.

101. Note, 90 mph$= 90 \cdot \dfrac{5280}{3600}$ ft/sec$= 132$ ft/sec.

In $v_o^2 \sin 2\theta = 32d$, let $v_o = 132$ and $d = 230$.

$$132^2 \sin 2\theta = 32(230)$$

$$\sin 2\theta = \frac{32(230)}{132^2}$$

$$\sin 2\theta \approx 0.4224$$

$$2\theta = \sin^{-1}(0.4224) \approx 25.0° \text{ or } 155°$$

$$\theta \approx 12.5° \quad \text{or} \quad 77.5°$$

The two possible angles are $12.5°$ and $77.5°$.
The time it takes the ball to reach home plate
can be found by using $x = v_o t \cos\theta$.
(See Example 11). For the angle $12.5°$, it takes

$$t = \frac{230}{132 \cos 12.5°} \approx 1.78 \text{ sec}$$

while for $77.5°$ it takes

$$t = \frac{230}{132 \cos 77.5°} \approx 8.05 \text{ sec}.$$

The difference in time is $8.05 - 1.78 \approx 6.3$ sec.

For Thought

1. True, for the sum of the measurements
of the three angles is $180°$.

2. False, $a \sin 17° = 88 \sin 9°$ and $a = \dfrac{88 \sin 9°}{\sin 17°}$.

3. False, since $\alpha = \sin^{-1}\left(\dfrac{5 \sin 44°}{18}\right) \approx 11°$ and
$\alpha = 180 - 11° = 169°$.

4. True

5. True, since $\dfrac{\sin 60°}{\sqrt{3}} = \dfrac{\sqrt{3}/2}{\sqrt{3}} = \dfrac{1}{2}$ and

$$\frac{\sin 30°}{1} = \sin 30° = \frac{1}{2}.$$

6. False, a triangle exists since $a = 500$ is bigger
than $h = 10 \sin 60° \approx 8.7$.

7. True

8. False, $a = \sqrt{c^2 + b^2 - 2bc \cos\alpha}$.

9. False, $c^2 = a^2 + b^2 - 2ab \cos\gamma$.

10. True

3.9 Exercises

1. Note $\gamma = 180° - (64° + 72°) = 44°$.

By the sine law $\dfrac{b}{\sin 72°} = \dfrac{13.6}{\sin 64°}$ and

$\dfrac{c}{\sin 44°} = \dfrac{13.6}{\sin 64°}$. So $b = \dfrac{13.6}{\sin 64°} \cdot \sin 72°$

≈ 14.4 and $c = \dfrac{13.6}{\sin 64°} \cdot \sin 44° \approx 10.5$.

3. Note $\beta = 180° - (12.2° + 33.6°) = 134.2°$.

By the sine law $\dfrac{a}{\sin 12.2°} = \dfrac{17.6}{\sin 134.2°}$

and $\dfrac{c}{\sin 33.6°} = \dfrac{17.6}{\sin 134.2°}$.

So $a = \dfrac{17.6}{\sin 134.2°} \cdot \sin 12.2° \approx 5.2$ and

$c = \dfrac{17.6}{\sin 134.2°} \cdot \sin 33.6° \approx 13.6$.

5. Note $\beta = 180° - (10.3° + 143.7°) = 26°$.

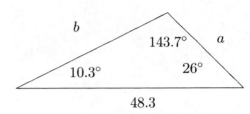

Since $\dfrac{a}{\sin 10.3°} = \dfrac{48.3}{\sin 143.7°}$ and

$\dfrac{b}{\sin 26°} = \dfrac{48.3}{\sin 143.7°}$, we have

$a = \dfrac{48.3}{\sin 143.7°} \cdot \sin 10.3° \approx 14.6$ and

$b = \dfrac{48.3}{\sin 143.7°} \cdot \sin 26° \approx 35.8$

7. Note $\alpha = 180° - (120.7° + 13.6°) = 45.7°$.

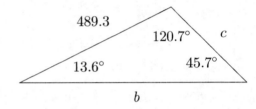

Since $\dfrac{c}{\sin 13.6°} = \dfrac{489.3}{\sin 45.7°}$ and

$\dfrac{b}{\sin 120.7°} = \dfrac{489.3}{\sin 45.7°}$, we have

$c = \dfrac{489.3}{\sin 45.7°} \cdot \sin 13.6° \approx 160.8$ and

$b = \dfrac{489.3}{\sin 45.7°} \cdot \sin 120.7° \approx 587.9$

9. Draw angle $\alpha = 39.6°$ and let h be the height.

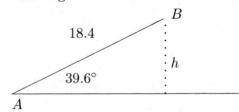

Since $\sin 39.6° = \dfrac{h}{18.4°}$, we have

$h = 18.4 \sin 39.6° \approx 11.7$. There is no triangle since $a = 3.7$ is smaller than $h \approx 11.7$.

11. Draw angle $\gamma = 60°$ and let h be the height.

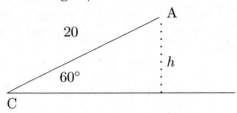

Since $h = 20 \sin 60° = 10\sqrt{3}$ and $c = h$, there is exactly one triangle and it is a right triangle. So $\beta = 90°$ and $\alpha = 30°$. By the Pythagorean Theorem,

$$a = \sqrt{20^2 - (10\sqrt{3})^2} = \sqrt{400 - 300} = 10.$$

13. Since β is an obtuse angle and $b > c$, there is exactly one triangle.

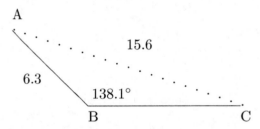

Apply the sine law.

$$\frac{15.6}{\sin 138.1°} = \frac{6.3}{\sin \gamma}$$

$$\sin \gamma = \frac{6.3 \sin 138.1°}{15.6}$$

$$\sin \gamma \approx 0.2697$$

$$\gamma = \sin^{-1}(0.2697) \approx 15.6°$$

So $\alpha = 180° - (15.6° + 138.1°) = 26.3°$. By the sine law, $a = \dfrac{15.6}{\sin 138.1°} \sin 26.3° \approx 10.3$.

15. Draw angle $\beta = 32.7°$ and let h be the height.

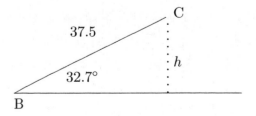

Since $h = 37.5 \sin 32.7° \approx 20.3$ and $20.3 < b < 37.5$, there are two triangles and they are given by

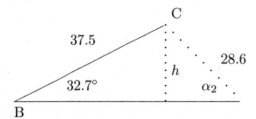

and

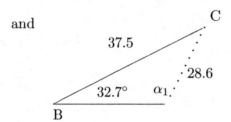

Apply the sine law to the acute triangle.

$$\frac{28.6}{\sin 32.7°} = \frac{37.5}{\sin \alpha_2}$$

$$\sin \alpha_2 = \frac{37.5 \sin 32.7°}{28.6}$$

$$\sin \alpha_2 \approx 0.708$$

$$\alpha_2 = \sin^{-1}(0.708) \approx 45.1°$$

So $\gamma_2 = 180° - (45.1° + 32.7°) = 102.2°$. By the sine law, $c_2 = \dfrac{28.6}{\sin 32.7°} \sin 102.2° \approx 51.7$.

On the obtuse triangle, we find $\alpha_1 = 180° - \alpha_2 = 134.9°$ and $\gamma_1 = 180° - (134.9° + 32.7°) = 12.4°$. By the sine law, $c_1 = \dfrac{28.6}{\sin 32.7°} \sin 12.4° \approx 11.4$.

17. Draw angle $\gamma = 99.6°$. Note, there is exactly one triangle since $12.4 > 10.3$.

By the sine law, we obtain

$$\frac{12.4}{\sin 99.6°} = \frac{10.3}{\sin \beta}$$

$$\sin \beta = \frac{10.3 \sin 99.6°}{12.4}$$

$$\sin \beta \approx 0.819$$

$$\beta = \sin^{-1}(0.819) \approx 55.0°.$$

So $\alpha = 180° - (55.0° + 99.6°) = 25.4°$.

By the sine law, $a = \dfrac{12.4}{\sin 99.6°} \sin 25.4° \approx 5.4$.

19. By the cosine law, we obtain

$$c = \sqrt{3.1^2 + 2.9^2 - 2(3.1)(2.9) \cos 121.3°}$$

$\approx 5.23 \approx 5.2$. By the sine law, we find

$$\frac{3.1}{\sin \alpha} = \frac{5.23}{\sin 121.3°}$$

$$\sin \alpha = \frac{3.1 \sin 121.3°}{5.23}$$

$$\sin \alpha \approx 0.50647$$

$$\alpha \approx \sin^{-1}(0.50647) \approx 30.4°.$$

Then $\beta = 180° - (30.4° + 121.3°) = 28.3°$.

21. By the cosine law, we find

$$\cos \beta = \frac{6.1^2 + 5.2^2 - 10.3^2}{2(6.1)(5.2)} \approx -0.6595$$

and so $\beta \approx \cos^{-1}(-0.6595) \approx 131.3°$.

By the sine law,

$$\frac{6.1}{\sin \alpha} = \frac{10.3}{\sin 131.3°}$$

$$\sin \alpha = \frac{6.1 \sin 131.3°}{10.3}$$

$$\sin \alpha \approx 0.4449$$

$$\alpha \approx \sin^{-1}(0.4449) \approx 26.4°.$$

So $\gamma = 180° - (26.4° + 131.3°) = 22.3°$.

23. By the cosine law,

$$b = \sqrt{2.4^2 + 6.8^2 - 2(2.4)(6.8) \cos 10.5°}$$

$\approx 4.46167 \approx 4.5$ and

$$\cos \alpha = \frac{2.4^2 + 4.46167^2 - 6.8^2}{2(2.4)(4.46167)} \approx -0.96066.$$

So $\alpha = \cos^{-1}(-0.96066) \approx 163.9°$ and $\gamma = 180° - (163.9° + 10.5°) = 5.6°$.

25. By the cosine law,

$$\cos\alpha = \frac{12.2^2 + 8.1^2 - 18.5^2}{2(12.2)(8.1)} \approx -0.6466.$$

Then $\alpha = \cos^{-1}(-0.6466) \approx 130.3°$.
By the sine law,

$$\frac{12.2}{\sin\beta} = \frac{18.5}{\sin 130.3°}$$

$$\sin\beta = \frac{12.2 \sin 130.3°}{18.5}$$

$$\sin\beta \approx 0.5029$$

$$\beta \approx \sin^{-1}(0.5029) \approx 30.2°$$

So $\gamma = 180° - (30.2° + 130.3°) = 19.5°$.

27. By the cosine law, we obtain

$$a = \sqrt{9.3^2 + 12.2^2 - 2(9.3)(12.2)\cos 30°}$$
$$\approx 6.23 \approx 6.2 \text{ and}$$
$$\cos\gamma = \frac{6.23^2 + 9.3^2 - 12.2^2}{2(6.23)(9.3)} \approx -0.203.$$

So $\gamma = \cos^{-1}(-0.203) \approx 101.7°$ and
$\beta = 180° - (101.7° + 30°) = 48.3°$.

29. By the cosine law,

$$\cos\beta = \frac{6.3^2 + 6.8^2 - 7.1^2}{2(6.3)(6.8)} \approx 0.4146.$$

So $\beta = \cos^{-1}(0.4146) \approx 65.5°$.
By the sine law, we have

$$\frac{6.8}{\sin\gamma} = \frac{7.1}{\sin 65.5°}$$

$$\sin\gamma = \frac{6.8 \sin 65.5°}{7.1}$$

$$\sin\gamma \approx 0.8715$$

$$\gamma \approx \sin^{-1}(0.8715) \approx 60.6°.$$

So $\alpha = 180° - (60.6° + 65.5°) = 53.9°$.

31. Note, $\alpha = 180° - 25° - 35° = 120°$.
Then by the sine law, we obtain

$$\frac{7.2}{\sin 120°} = \frac{b}{\sin 25°} = \frac{c}{\sin 35°}$$

from which we have

$$b = \frac{7.2 \sin 25°}{\sin 120°} \approx 3.5$$

and

$$c = \frac{7.2 \sin 35°}{\sin 120°} \approx 4.8.$$

33. There is no such triangle. Note,

$$a + b = c$$

and in a triangle the sum of the lengths of two sides is greater than the length of the third side.

35. One triangle exists. The angles are uniquely determined by the law of cosines.

37. There is no such triangle since the sum of the angles in a triangle is 180°.

39. Exactly one triangle exists. This is seen by constructing a 179°-angle with two sides that have lengths 1 and 10. The third side is constructed by joining the endpoints of the first two sides.

41. Consider the figure below.

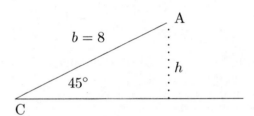

Note,

$$h = 8 \sin 45° = 2\sqrt{2}.$$

So the minimum value of c so that we will be able to make a triangle is $2\sqrt{2}$. Since $c = 2$, no such triangle is possible.

43. Let x be the number of miles flown along I-20.

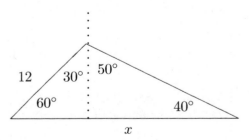

By the sine law, $\dfrac{x}{\sin 80°} = \dfrac{12}{\sin 40°}$.
Then $x \approx 18.4$ miles.

45. Let h be the height of the tower.

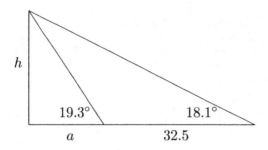

By using right triangle trigonometry, we get $\tan 19.3° = \dfrac{h}{a}$ or $a = \dfrac{h}{\tan 19.3°}$. Similarly, we have $\tan 18.1° = \dfrac{h}{a + 32.5}$. Then

$$\tan 18.1°(a + 32.5) = h$$
$$a\tan 18.1° + 32.5\tan 18.1° = h$$
$$\frac{h}{\tan 19.3°} \cdot \tan 18.1° + 32.5\tan 18.1° = h$$
$$h \cdot \frac{\tan 18.1°}{\tan 19.3°} + 32.5\tan 18.1° = h.$$

Solving for h, we find that the height of the tower is $h \approx 159.4$ ft.

47. Note, $\tan\gamma = 6/12$ and $\gamma = \tan^{-1}(0.5) \approx 26.565°$. Also, $\tan\alpha = 3/12$ and $\alpha = \tan^{-1}(0.25) \approx 14.036°$.

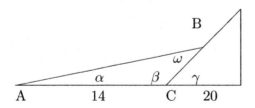

The remaining angles are $\beta = 153.435°$ and $\omega = 12.529°$.

By the sine law, $\dfrac{AB}{\sin 153.435°} = \dfrac{14}{\sin 12.529°}$

and $\dfrac{BC}{\sin 14.036°} = \dfrac{14}{\sin 12.529°}$.

Then $AB \approx 28.9$ ft and $BC \approx 15.7$ ft.

49. Recall, a central angle α in a circle of radius r intercepts a chord of length $r\sqrt{2 - 2\cos\alpha}$. Since $r = 30$ and $\alpha = 19°$, the length is

$$30\sqrt{2 - 2\cos 19°} \approx 9.90 \text{ ft.}$$

51. Note, a central angle α in a circle of radius r intercepts a chord of length $r\sqrt{2 - 2\cos\alpha}$. Since $921 = r\sqrt{2 - 2\cos 72°}$ (where $360 \div 5 = 72$), we get

$$r = \frac{921}{\sqrt{2 - 2\cos 72°}} \approx 783.45 \text{ ft.}$$

53. After 6 hours, Jan hiked a distance of 24 miles and Dean hiked 30 miles. Let x be the distance between them after 6 hrs.

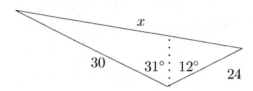

By the cosine law,
$$x = \sqrt{30^2 + 24^2 - 2(30)(24)\cos 43°} =$$
$$\sqrt{1476 - 1440\cos 43°} \approx 20.6 \text{ miles.}$$

55. Consider the figure below.

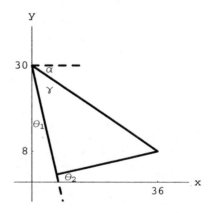

Note $\tan\alpha = 22/36$ and

$$\alpha = \tan^{-1}(22/36) \approx 31.4°.$$

The distance between $(36, 8)$ and $(0, 30)$ is approximately 42.19. By the cosine law,

$$\cos\beta = \frac{30^2 + 30^2 - 42.19^2}{2(30)(30)}$$
$$\cos\beta \approx 0.011$$
$$\beta \approx \cos^{-1}(0.011) \approx 89.4°.$$

So, we obtain

$$\theta_2 = 180° - 89.4° = 90.6°.$$

By the sine law, we find

$$\frac{30}{\sin\gamma} = \frac{42.19}{\sin 89.4°}$$

$$\sin\gamma \approx 0.711$$

$$\gamma \approx \sin^{-1}(0.711) \approx 45.3°.$$

Then we find

$$\theta_1 = 90° - (45.3° + 31.4°) = 13.3°.$$

Chapter 3 Review Exercises

1. $388° - 360° = 28°$

3. $-153°14'27'' + 359°59'60'' = 206°45'33''$

5. $180°$

7. $13\pi/5 - 2\pi = 3\pi/5 = 3 \cdot 36° = 108°$

9. $5\pi/3 = 5 \cdot 60° = 300°$

11. $270°$ **13.** $11\pi/6$ **15.** $-5\pi/3$

17.

θ deg	0	30	45	60	90	120	135	150	180
θ rad	0	$\frac{\pi}{6}$	$\frac{\pi}{4}$	$\frac{\pi}{3}$	$\frac{\pi}{2}$	$\frac{2\pi}{3}$	$\frac{3\pi}{4}$	$\frac{5\pi}{6}$	π
$\sin\theta$	0	$\frac{1}{2}$	$\frac{\sqrt{2}}{2}$	$\frac{\sqrt{3}}{2}$	1	$\frac{\sqrt{3}}{2}$	$\frac{\sqrt{2}}{2}$	$\frac{1}{2}$	0
$\cos\theta$	1	$\frac{\sqrt{3}}{2}$	$\frac{\sqrt{2}}{2}$	$\frac{1}{2}$	0	$-\frac{1}{2}$	$-\frac{\sqrt{2}}{2}$	$-\frac{\sqrt{3}}{2}$	-1
$\tan\theta$	0	$\frac{\sqrt{3}}{3}$	1	$\sqrt{3}$	NA	$-\sqrt{3}$	-1	$-\frac{\sqrt{3}}{3}$	0

19. $-\sqrt{2}/2$ **21.** $\sqrt{3}$ **23.** $-2\sqrt{3}/3$

25. 0 **27.** 0

29. -1 **31.** $\cot(60°) = \sqrt{3}/3$ **33.** $-\sqrt{2}/2$

35. -2 **37.** $-\sqrt{3}/3$

39. $\sin(\alpha) = 5/13, \cos(\alpha) = 12/13, \tan(\alpha) = 5/12,$
$\csc(\alpha) = 13/5, \sec(\alpha) = 13/12, \cot(\alpha) = 12/5$

41. 0.6947 **43.** -0.0923 **45.** 0.1869

47. 1.0356 **49.** $-\pi/6$

51. $-\pi/4$ **53.** $\pi/4$

54. $\pi/6$ **55.** $\pi/6$

57. $90°$ **59.** $135°$

61. $30°$ **63.** $90°$

65. Form the right triangle with $a = 2$, $b = 3$.

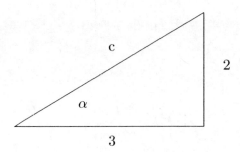

Note that $c = \sqrt{2^2 + 3^2} = \sqrt{13}$, $\tan(\alpha) = 2/3$, so $\alpha = \tan^{-1}(2/3) \approx 33.7°$ and $\beta \approx 56.3°$.

67. Form the right triangle with $a = 3.2$, $\alpha = 21.3°$.

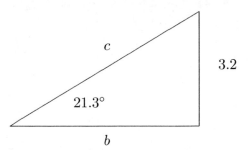

Since $\sin 21.3° = \dfrac{3.2}{c}$ and

$\tan 21.3° = \dfrac{3.2}{b}$, $c = \dfrac{3.2}{\sin 21.3°} \approx 8.8$

and $b = \dfrac{3.2}{\tan 21.3°} \approx 8.2$

Also, $\beta = 90° - 21.3° = 68.7°$

69. $f(x) = 2\sin(3x)$ has period $2\pi/3$, range $[-2, 2]$

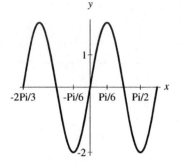

71. $f(x) = \tan(2x + \pi)$ has period $\pi/2$, range $(-\infty, \infty)$

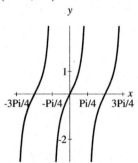

73. $f(x) = \sec(x/2)$ has period 4π, range $(-\infty, -1] \cup [1, \infty)$

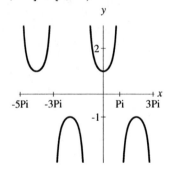

75. $f(x) = \dfrac{1}{2} \cdot \cos(2x)$ has period π, range $[-1/2, 1/2]$

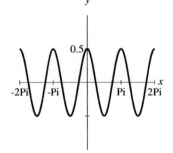

77. $f(x) = \cot\left(2x + \dfrac{\pi}{3}\right)$ has period $\dfrac{\pi}{2}$, range $(-\infty, \infty)$

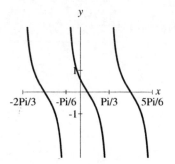

79. $f(x) = \dfrac{1}{3} \cdot \csc(2x + \pi)$ has period π, range $(-\infty, -1/3] \cup [1/3, \infty)$

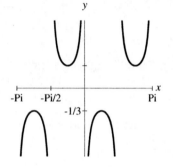

81. One full period is between $2 \le x \le 6$.

So $\dfrac{2\pi}{b} = 4$ or $b = \dfrac{\pi}{2}$ and there is a right shift of 2 units. An equation is $y = 2\sin\left(\dfrac{\pi}{2}(x - 2)\right)$.

83. Note, one full period is between $8\pi \le x \le 10\pi$, there is a vertical upward shift of 40 units, and $a = 20$. An equation is

$$y = 20\sin(x) + 40.$$

85. $\alpha = 150°$

87. $\sin(\alpha) = -\sqrt{1 - (1/5)^2} = -\sqrt{24/25} = -2\sqrt{6}/5$

89. In the given right triangle, the hypotenuse is 24 feet and the side adjacent to the $16°$ angle is 18 feet. Since $\sin(16°) = \dfrac{x}{24}$, the shortest side is

$$x = 24\sin(16°) \approx 6.6 \text{ feet.}$$

91. Since the largest angle α must be opposite the 8 cm leg (which is the longest leg) and the leg adjacent to α is 6 cm long, we obtain

$$\alpha = \tan^{-1}\left(\frac{8}{6}\right) \approx 53.1°.$$

93. $1 - \sin^2\alpha = \cos^2\alpha$

95. $(1 - \csc x)(1 + \csc x) = 1 - \csc^2 x = -\cot^2 x$

97.

$$\frac{1}{1 + \sin\alpha} + \frac{\sin\alpha}{\cos^2\alpha} = \frac{1}{1 + \sin\alpha} + \frac{\sin\alpha}{1 - \sin^2\alpha} =$$
$$\frac{(1 - \sin\alpha) + \sin\alpha}{1 - \sin^2\alpha} = \frac{1}{\cos^2\alpha} = \sec^2\alpha$$

99 $\tan(4s)$, by the double angle idenity for tangent

101. $\sin(3\theta - 6\theta) = \sin(-3\theta) = -\sin(3\theta)$

103. $\tan\left(\dfrac{2z}{2}\right) = \tan z$, by a double-angle identity

for tangent

105.

$$= \frac{1 + \tan^2\theta}{1 - \tan^2\theta}$$

$$= \frac{\sec^2\theta}{1 - \dfrac{\sin^2\theta}{\cos^2\theta}} \cdot \frac{\cos^2\theta}{\cos^2\theta}$$

$$= \frac{1}{\cos^2\theta - \sin^2\theta}$$

$$= \frac{1}{\cos 2\theta}$$

$$= \sec 2\theta$$

107.

$$= \frac{\csc^2 x - \cot^2 x}{2\csc^2 x + 2\csc x \cot x}$$

$$= \frac{1}{\dfrac{2}{\sin^2 x} + 2 \cdot \dfrac{1}{\sin x} \cdot \dfrac{\cos x}{\sin x}}$$

$$= \frac{1}{\dfrac{2}{\sin^2 x} + \dfrac{2\cos x}{\sin^2 x}} \cdot \frac{\sin^2 x}{\sin^2 x}$$

$$= \frac{\sin^2 x}{2 + 2\cos x}$$

$$= \frac{1 - \cos^2 x}{2(1 + \cos x)}$$

$$= \frac{(1 - \cos x)(1 + \cos x)}{2(1 + \cos x)}$$

$$= \frac{1 - \cos x}{2}$$

$$= \sin^2\left(\frac{x}{2}\right)$$

109.

$$\cot(\alpha - 45°) =$$
$$(\tan(\alpha - 45°))^{-1} =$$
$$\left(\frac{\tan\alpha - \tan 45°}{1 + \tan\alpha \tan 45°}\right)^{-1} =$$
$$\left(\frac{\tan\alpha - 1}{1 + \tan\alpha}\right)^{-1} =$$
$$\frac{1 + \tan\alpha}{\tan\alpha - 1} =$$

111.

$$\frac{\sin 2\beta}{2\csc\beta} =$$
$$\frac{2\sin\beta\cos\beta}{2/\sin\beta} \cdot \frac{\sin\beta}{\sin\beta} =$$
$$\sin^2\beta\cos\beta =$$

113. Factor the numerator on the left-hand side as a difference of two cubes.
Note, $\cot w \tan w = 1$.

$$\frac{\cot^3 y - \tan^3 y}{\sec^2 y + \cot^2 y} =$$
$$\frac{(\cot y - \tan y)(\cot^2 y + 1 + \tan^2 y)}{\sec^2 y + \cot^2 y} =$$
$$\frac{(\cot y - \tan y)(\cot^2 y + \sec^2 y)}{\sec^2 y + \cot^2 y} =$$
$$\cot y - \tan y =$$
$$\frac{1}{\tan y} - \tan y =$$
$$\frac{1 - \tan^2 y}{\tan y} =$$

$$2 \cdot \frac{1 - \tan^2 y}{2 \tan y} \quad =$$

$$2 \cdot (\tan 2y)^{-1} \quad =$$

$$2 \cot(2y) \quad =$$

115. Isolate $\cos 2x$ on one side.

$$2 \cos 2x \quad = \quad -1$$

$$\cos 2x \quad = \quad -\frac{1}{2}$$

$$2x = \frac{2\pi}{3} + 2k\pi \quad \text{or} \quad 2x = \frac{4\pi}{3} + 2k\pi$$

$$x = \frac{\pi}{3} + k\pi \quad \text{or} \quad x = \frac{2\pi}{3} + k\pi$$

The solution set is

$$\left\{ x \mid x = \frac{\pi}{3} + k\pi \text{ or } x = \frac{2\pi}{3} + k\pi \right\}.$$

117. Set each factor to zero.

$$(\sqrt{3} \csc x - 2)(\csc x - 2) \quad = \quad 0$$

$$\csc x = \frac{2}{\sqrt{3}} \quad \text{or} \quad \csc x = 2$$

Thus, $x = \frac{\pi}{3}, \frac{2\pi}{3}, \frac{\pi}{6}, \frac{5\pi}{6}$ plus multiples of 2π.
The solution set is

$$\left\{ x \mid x = \frac{\pi}{3} + 2k\pi, \frac{2\pi}{3} + 2k\pi, \frac{\pi}{6} + 2k\pi, \frac{5\pi}{6} + 2k\pi \right\}.$$

119. Set the right-hand side to zero and factor.

$$2 \sin^2 x - 3 \sin x + 1 \quad = \quad 0$$

$$(2 \sin x - 1)(\sin x - 1) \quad = \quad 0$$

$$\sin x = \frac{1}{2} \quad \text{or} \quad \sin x = 1$$

The $x = \frac{\pi}{6}, \frac{5\pi}{6}, \frac{\pi}{2}$ plus multiples of 2π.
The solution set is

$$\left\{ x \mid x = \frac{\pi}{6} + 2k\pi, \frac{5\pi}{6} + 2k\pi, \frac{\pi}{2} + 2k\pi \right\}.$$

121. Isolate $\sin \frac{x}{2}$ on one side.

$$\sin \frac{x}{2} \quad = \quad \frac{12}{8\sqrt{3}}$$

$$\sin \frac{x}{2} \quad = \quad \frac{3}{2\sqrt{3}}$$

$$\sin \frac{x}{2} \quad = \quad \frac{\sqrt{3}}{2}$$

$$\frac{x}{2} = \frac{\pi}{3} + 2k\pi \quad \text{or} \quad \frac{x}{2} = \frac{2\pi}{3} + 2k\pi$$

$$x = \frac{2\pi}{3} + 4k\pi \quad \text{or} \quad x = \frac{4\pi}{3} + 4k\pi$$

The solution set is

$$\left\{ x \mid x = \frac{2\pi}{3} + 4k\pi \text{ or } x = \frac{4\pi}{3} + 4k\pi \right\}.$$

123. By using the double-angle identity for sine, we get

$$\cos \frac{x}{2} - \sin \left(2 \cdot \frac{x}{2} \right) \quad = \quad 0$$

$$\cos \frac{x}{2} - 2 \sin \frac{x}{2} \cos \frac{x}{2} \quad = \quad 0$$

$$\cos \frac{x}{2} \left(1 - 2 \sin \frac{x}{2} \right) \quad = \quad 0$$

$$\cos \frac{x}{2} = 0 \quad \text{or} \quad \sin \frac{x}{2} = \frac{1}{2}.$$

Then $\frac{x}{2} = \frac{\pi}{2}, \frac{\pi}{6}, \frac{5\pi}{6}$ plus mutiples of 2π.

Or $x = \pi, \frac{\pi}{3}, \frac{5\pi}{3}$ plus multiples of 4π.
The solution set is

$$\left\{ x \mid x = \pi + 4k\pi, \frac{\pi}{3} + 4k\pi, \frac{5\pi}{3} + 4k\pi \right\}.$$

125. By the double-angle identity for cosine, we find

$$\cos 2x + \sin^2 x \quad = \quad 0$$

$$\cos^2 x - \sin^2 x + \sin^2 x \quad = \quad 0$$

$$\cos^2 x \quad = \quad 0$$

$$x \quad = \quad \frac{\pi}{2} + k\pi.$$

The solution set is $\left\{ x \mid x = \frac{\pi}{2} + k\pi \right\}.$

127. Draw a triangle with $\gamma = 48°$, $a = 3.4$, and $b = 2.6$.

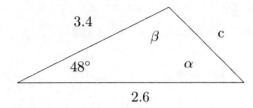

By the cosine law, we obtain

$c = \sqrt{2.6^2 + 3.4^2 - 2(2.6)(3.4)\cos 48°} \approx$

$2.5475 \approx 2.5$. By the sine law, we find

$$\frac{2.5475}{\sin 48°} = \frac{2.6}{\sin \beta}$$

$$\sin \beta = \frac{2.6 \sin 48°}{2.5}$$

$$\sin \beta \approx 0.75846$$

$$\beta \approx \sin^{-1}(0.75846)$$

$$\beta \approx 49.3°.$$

Also, $\alpha = 180° - (49.3° + 48°) = 82.7°$.

129. Draw a triangle with $\alpha = 13°$, $\beta = 64°$, and $c = 20$.

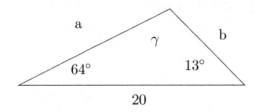

Note $\gamma = 180° - (64° + 13°) = 103°$.

By the sine law, we get

$$\frac{20}{\sin 103°} = \frac{a}{\sin 13°}$$

and

$$\frac{20}{\sin 103°} = \frac{b}{\sin 64°}.$$

So

$$a = \frac{20}{\sin 103°} \sin 13° \approx 4.6$$

and

$$b = \frac{20}{\sin 103°} \sin 64° \approx 18.4.$$

131. Draw a triangle with $a = 3.6$, $b = 10.2$, and $c = 5.9$.

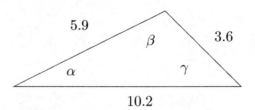

By the cosine law one gets

$$\cos \beta = \frac{5.9^2 + 3.6^2 - 10.2^2}{2(5.9)(3.6)} \approx -1.3.$$

This is a contradiction since the range of cosine is $[-1, 1]$. No triangle exists.

133. Draw a triangle with sides $a = 30.6$, $b = 12.9$, and $c = 24.1$.

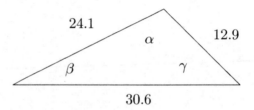

By the cosine law, we get

$$\cos \alpha = \frac{24.1^2 + 12.9^2 - 30.6^2}{2(24.1)(12.9)} \approx -0.3042.$$

So, we obtain

$$\alpha = \cos^{-1}(-0.3042) \approx 107.7°.$$

Similarly, we find

$$\cos \beta = \frac{24.1^2 + 30.6^2 - 12.9^2}{2(24.1)(30.6)} \approx 0.9158.$$

So, we find

$$\beta = \cos^{-1}(0.9158) \approx 23.7°$$

and

$$\gamma = 180° - (107.7° + 23.7°) = 48.6°.$$

135. Draw angle $\beta = 22°$ and let h be the height.

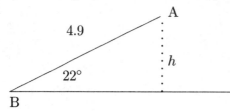

Since $h = 4.9 \sin 22° \approx 1.8$ and $1.8 < b < 4.9$, we have two triangles and they are given by

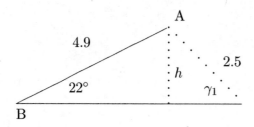

and

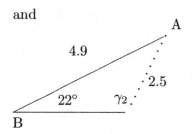

Apply the sine law to case 1.

$$\frac{4.9}{\sin \gamma_1} = \frac{2.5}{\sin 22°}$$

$$\sin \gamma_1 = \frac{4.9 \sin 22°}{2.5}$$

$$\sin \gamma_1 \approx 0.7342$$

$$\gamma_1 = \sin^{-1}(0.7342) \approx 47.2°$$

So $\alpha_1 = 180° - (22° + 47.2°) = 110.8°$.

By the sine law, $a_1 = \dfrac{2.5}{\sin 22°} \sin 110.8° \approx 6.2$.

In case 2, $\gamma_2 = 180° - \gamma_1 = 132.8°$ and $\alpha_2 = 180° - (22° + 132.8°) = 25.2°$.

By the sine law, $a_2 = \dfrac{2.5}{\sin 22°} \sin 25.2° \approx 2.8$.

137. Period is $\dfrac{1}{92.3 \times 10^6} \approx 1.08 \times 10^{-8}$ sec

139. The height of the man is

$$s = r \cdot \alpha = 1000(0.4) \cdot \frac{\pi}{180} \approx 6.9813 \text{ ft.}$$

141. Since the period is 20 minutes, $\dfrac{2\pi}{b} = 20$ or $b = \dfrac{\pi}{10}$. Since the depth is between 12 ft and 16 ft, the vertical upward shift is 14 and $a = 2$. Since the depth is 16 ft at time $t = 0$, one can assume there is a left shift of 5 minutes. An equation is

$$y = 2 \sin\left(\frac{\pi}{10}(x + 5) \right) + 14$$

and its graph is given.

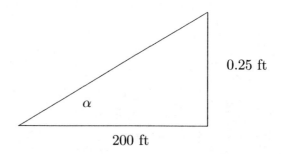

143. Form the right triangle below.

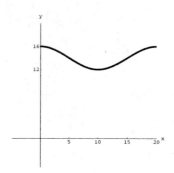

Since $\tan(\alpha) = 0.25/200$, we find

$$\alpha = \tan^{-1}(0.25/200) \approx 0.0716°.$$

She will not hit the target if she deviates by 0.1° from the center of the circle.

145. Let h be the height of the building.

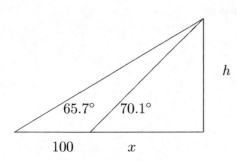

Since $\tan 70.1° = \dfrac{h}{x}$ and $\tan 65.7° = \dfrac{h}{100 + x}$, we obtain

$$\tan(65.7°) = \frac{h}{100 + h/\tan(70.1°)}$$

$$100 \cdot \tan(65.7°) + h \cdot \frac{\tan(65.7°)}{\tan(70.1°)} = h$$

$$100 \cdot \tan(65.7°) = h \left(1 - \frac{\tan(65.7°)}{\tan(70.1°)}\right)$$

$$h = \frac{100 \cdot \tan(65.7°)}{1 - \tan(65.7°)/\tan(70.1°)}$$

$$h \approx 1117 \text{ ft.}$$

147. Consider the right triangle shown where vertex A represents the center of the earth, the cargo ship is located at vertex B, and d is the distance to the horizon.

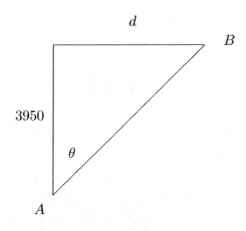

From the triangle one finds

$$\tan \theta = \frac{d}{3950}$$

and since

$$AB = 3950 + \frac{90}{5280}$$

one obtains

$$\sin \theta = \frac{d}{3950 + \frac{90}{5280}}.$$

From the first identity, one can derive

$$\sin \theta = \frac{d}{\sqrt{d^2 + 3950^2}}.$$

Then the distance to the horizon is given by

$$\frac{d}{\sqrt{d^2 + 3950^2}} = \frac{d}{3950 + \frac{90}{5280}}$$

$$\frac{1}{\sqrt{d^2 + 3950^2}} = \frac{1}{3950 + \frac{90}{5280}}$$

$$3950 + \frac{90}{5280} = \sqrt{d^2 + 3950^2}$$

$$\left(3950 + \frac{90}{5280}\right)^2 = d^2 + 3950^2$$

$$\left(3950 + \frac{90}{5280}\right)^2 - 3950^2 = d^2$$

$$\sqrt{\left(3950 + \frac{90}{5280}\right)^2 - 3950^2} = d$$

$$11.6 \text{ miles} = d.$$

For Thought

1. False, the base of an exponential function is a positive number and -2 is a negative number.

2. True

3. True

4. True   **5.**   True   **6.**   True   **7.**   True

8. False, since it is decreasing.

9. True, since $0.25 = 4^{-1}$.

10. True, since $\sqrt[100]{2^{173}} = \left(2^{173}\right)^{1/100}$.

4.1 Exercises

1. $3^2 = 9$

3. $3^{-2} = 1/9$

5. $2^{-1} = 1/2$

7. $2^3 = 8$

9. $(1/4)^{-1} = 4$

11. $4^{1/2} = 2$

13. $f(x) = 5^x$ goes through $(-1, 1/5), (0, 1), (1, 5)$, domain is $(-\infty, \infty)$, range is $(0, \infty)$, increasing

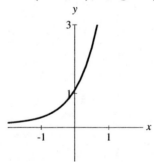

15. $f(x) = 10^{-x}$ goes through $(-1, 10), (0, 1)$, $(1, 1/10)$, domain is $(-\infty, \infty)$, range is $(0, \infty)$, decreasing

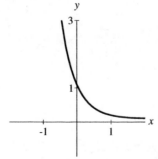

17. $f(x) = (1/4)^x$ goes through $(-1, 4), (0, 1)$, $(1, 1/4)$, domain is $(-\infty, \infty)$, range is $(0, \infty)$, decreasing

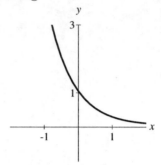

19. From the graph, we find $\lim\limits_{x \to \infty} 3^x = \infty$.

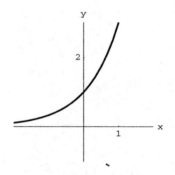

21. Using the graph, we obtain $\lim\limits_{x \to \infty} 5^{-x} = 0$.

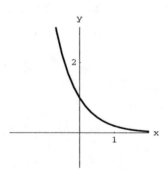

23. We see from the graph that $\lim\limits_{x \to \infty} \left(\dfrac{1}{3}\right)^x = 0$.

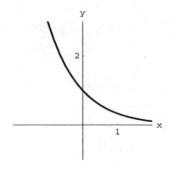

25. Using the graph, we get $\lim\limits_{x \to -\infty} e^{-x} = \infty$.

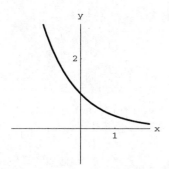

27. Shift $y = 2^x$ down by 3 units; $f(x) = 2^x - 3$ goes through $(-1, -2.5), (0, -2), (2, 1)$, domain $(-\infty, \infty)$, range $(-3, \infty)$, asymptote $y = -3$, increasing

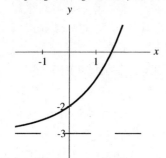

29. Shift $y = 2^x$ to left by 3 units and down by 5 units; $f(x) = 2^{x+3} - 5$ goes through $(-4, -4.5), (-3, -4), (0, 3)$, domain $(-\infty, \infty)$, range $(-5, \infty)$, asymptote $y = -5$, increasing

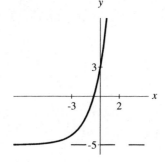

31. Reflect $y = 2^{-x}$ about x-axis, $f(x) = -2^{-x}$ goes through $(-1, -2), (0, -1), (1, -1/2)$, domain $(-\infty, \infty)$, range $(-\infty, 0)$, asymptote $y = 0$, increasing

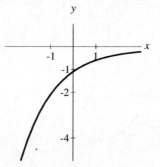

33. Reflect $y = 2^x$ about x-axis and shift up by 1 unit, $f(x) = 1 - 2^x$ goes through $(-1, 0.5), (0, 0), (1, -1)$, domain $(-\infty, \infty)$, range $(-\infty, 1)$, asymptote $y = 1$, decreasing

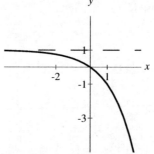

35. Shift $y = 3^x$ to right by 2 and shrink by a factor of 0.5, $f(x) = 0.5 \cdot 3^{x-2}$ goes through $(0, 1/18), (2, 0.5), (3, 1.5)$, domain $(-\infty, \infty)$, range $(0, \infty)$, asymptote $y = 0$, increasing

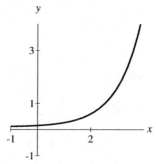

37. Stretch $y = (0.5)^x$ by a factor of 500,
$f(x) = 500 \cdot (0.5)^x$ goes through $(0, 500)$,
$(1, 250)$, domain $(-\infty, \infty)$, range $(0, \infty)$,
asymptote $y = 0$, decreasing

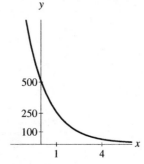

39. $y = 2^{x-5} - 2$

41. $y = -\left(\dfrac{1}{4}\right)^{x-1} - 2$

43. Since $2^x = 2^6$, solution set is $\{6\}$.

45. $\{-1\}$

47. Multiplying the equation by -1, $3^x = 3^3$
and the solution set is $\{3\}$.

49. $\{-2\}$

51. Since $(2^3)^x = 2^{3x} = 2$, $3x = 1$.
The solution set is $\left\{\dfrac{1}{3}\right\}$.

53. $\{-2\}$

55. Since $(2^{-1})^x = 2^{-x} = 2^3$, $-x = 3$.
The solution set is $\{-3\}$.

57. Since $10^{x-1} = 10^{-2}$, $x - 1 = -2$.
The solution set is $\{-1\}$.

59. Since $2^x = 4$, we find $x = 2$.

61. Since $2^x = \dfrac{1}{2}$, we obtain $x = -1$.

63. Since $\left(\dfrac{1}{3}\right)^x = 1$, we get $x = 0$.

65. Since $\left(\dfrac{1}{3}\right)^x = 3^3$, we find $x = -3$.

67. Since $10^x = 1000$, we obtain $x = 3$.

69. Since $10^x = 0.1 = 10^{-1}$, we find $x = -1$.

71. 1   **73.** -1

75. $(2, 9), (1, 3), (-1, 1/3), (-2, 1/9)$

77. $(0, 1), (-2, 25), (-1, 5), (1, 1/5)$

79. $(4, -16), (-2, -1/4), (-1, -1/2), (5, -32)$

81. When interest is compounded n times a year,
the amount at the end of 6 years is

$$A(n) = 5000\left(1 + \frac{0.08}{n}\right)^{6n}$$

and the interest earned after 6 years is

$$I(n) = A(n) - 5000.$$

a) If $n = 1$, then $A(1) = \$7934.37$ and
$I(1) = \$2934.37$.

b) If $n = 4$, then $A(4) = \$8042.19$ and
$I(4) = \$3042.19$.

c) If $n = 12$, then $A(12) = \$8067.51$ and
$I(12) = \$3067.51$.

d) If $n = 365$, then $A(365) = \$8079.95$ and
$I(365) = \$3079.95$.

83. After t years, a deposit of \$5000 will amount
to

$$A(t) = 5000e^{0.08t}.$$

We use 30 days per month and 365 days per
year.

a) After 6 years, the amount is
$A(6) = \$8080.37$.

b) After 8 years and 3 months, the amount is
$A\left(8 + \dfrac{90}{365}\right) = \$9671.31$.

c) After 5 years, 4 months, and 22 days, the
amount is

$$A\left(8 + \frac{4(30) + 22}{365}\right) = \$7694.93.$$

d) After 20 years and 321 days, the amount
is

$$A\left(20 + \frac{321}{365}\right) = \$26,570.30.$$

85. Assume there are 365 days in a year and 30 days in a month. The present value is

$$3000\left(1+\frac{0.065}{365}\right)^{-(365)(5+120/365)} = \$2121.82.$$

87. $20,000e^{-0.0542(30)} = \$3934.30$

89. a) The interest for first hour is

$$10^6 \cdot e^{0.06\left(\frac{1}{(24)(365)}\right)} - 10^6 = \$6.85.$$

b) The interest for 500th hour is the difference between the amounts in the account at the end of the 500th and 499th hours i.e.

$$10^6 e^{0.06\left(\frac{500}{(24)(365)}\right)} - 10^6 e^{0.06\left(\frac{499}{(24)(365)}\right)} = \$6.87$$

91. If $t = 0$, then $A = 200e^0 = 200$ g.
If $t = 500$, then $A = 200e^{-0.001(500)} \approx 121.3$ g.

93. $P = 10\left(\dfrac{1}{2}\right)^n$

95. When $t = 31$, the number of damaged O-rings is $n = 644e^{-0.15(31)} \approx 6$.

For Thought

1. True   **2.**   False, since $\log_{100}(10) = 1/2$.

3. True   **4.**   True

5. False, the domain is $(0, \infty)$.   **6.**   True

7. True

8. False, since $\log_a(0)$ is undefined.

9. True   **10.**   True

4.2 Exercises

1. 6   **3.** -4

5. $\dfrac{1}{4}$ **7.** -3

9. Since $2^6 = 64$, $\log_2(64) = 6$.

11. Since $3^{-4} = \dfrac{1}{81}$, $\log_3\left(\dfrac{1}{81}\right) = -4$.

13. Since $16^{1/4} = 2$, $\log_{16}(2) = \dfrac{1}{4}$.

15. Since $\left(\dfrac{1}{5}\right)^{-3} = 125$, $\log_{1/5}(125) = -3$.

17. Since $10^{-1} = 0.1$, $\log(0.1) = -1$

19. Since $10^0 = 1$, $\log(1) = 0$.

21. Since $e^1 = e$, $\ln(e) = 1$.

23. -5

25. $y = \log_3(x)$ goes through $(1/3, -1)$, $(1, 0)$, $(3, 1)$, domain $(0, \infty)$, range $(-\infty, \infty)$

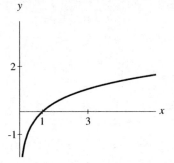

27. $y = \log_{1/2}(x)$ goes through $(2, -1), (1, 0)$, $(1/2, 1)$, domain $(0, \infty)$, range $(-\infty, \infty)$

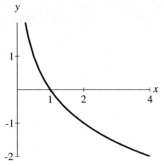

29. $f(x) = \ln(x - 1)$ goes through $\left(1 + \dfrac{1}{e}, -1\right)$, $(2, 0)$, $(1 + e, 1)$, domain $(1, \infty)$, range $(-\infty, \infty)$

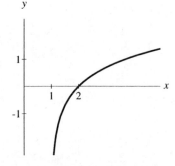

31. $f(x) = -3 + \log(x + 2)$ goes through
$(-1.9, -4)$, $(-1, -3)$, $(8, -2)$,
domain $(-2, \infty)$, range $(-\infty, \infty)$

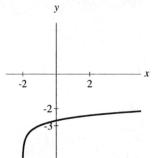

33. $f(x) = -\dfrac{1}{2}\log(x - 1)$ goes through $(1.1, 0.5)$,

$(2, 0)$, $(11, -0.5)$, domain $(1, \infty)$,
range $(-\infty, \infty)$

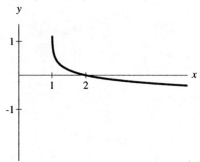

35. From the graph, we find $\lim\limits_{x \to \infty} \log_3 x = \infty$.

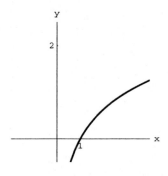

37. Using the graph, we obtain $\lim\limits_{x \to 0^+} \log_{1/2} x = \infty$.

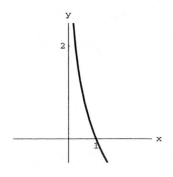

39. From the graph, we get $\lim\limits_{x \to 0^+} \ln x = -\infty$.

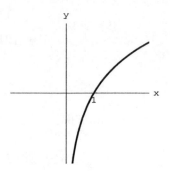

41. From the graph, we find $\lim\limits_{x \to \infty} \log x = \infty$.

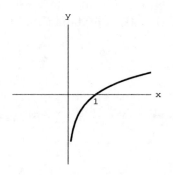

43. $y = \ln(x - 3) - 4$

45. $y = -\log_2(x - 5) - 1$

47. $2^5 = 32$

49. $5^y = x$

51. $10^z = 1000$

53. $\log_5(125) = 3$

55. $\ln(y) = 3$

57. $\log y = m$

59. $f^{-1}(x) = \log_2(x)$

61. $f^{-1}(x) = 7^x$

63. Replace $f(x)$ by y, interchange x and y,
solve for y, and replace y by $f^{-1}(x)$.

$$
\begin{aligned}
y &= \ln(x - 1) \\
x &= \ln(y - 1) \\
e^x &= y - 1 \\
y = f^{-1}(x) &= e^x + 1
\end{aligned}
$$

65. Replace $f(x)$ by y, interchange x and y, solve for y, and replace y by $f^{-1}(x)$.

$$
\begin{aligned}
y &= 3^{x+2} \\
x &= 3^{y+2} \\
y + 2 &= \log_3(x) \\
y = f^{-1}(x) &= \log_3(x) - 2
\end{aligned}
$$

67. Replace $f(x)$ by y, interchange x and y, solve for y, and replace y by $f^{-1}(x)$.

$$
\begin{aligned}
x &= \frac{1}{2}10^{y-1} + 5 \\
x - 5 &= \frac{1}{2}10^{y-1} \\
2x - 10 &= 10^{y-1} \\
\log(2x - 10) &= y - 1 \\
\log(2x - 10) + 1 &= y \\
f(x)^{-1} &= \log(2x - 10) + 1
\end{aligned}
$$

69. Since $2^8 = x$, the solution is $x = 256$.

71. Since $3^{1/2} = x$, the solution is $x = \sqrt{3}$.

73. Since $x^2 = 16$ and the base of a logarithm is positive, $x = 4$.

75. By using the definition of the logarithm, we get
$$x = \log_3(77).$$

77. Since $y = \ln(x)$ is one-to-one, we conclude that
$$x - 3 = 2x - 9.$$
Thus, the solution is $x = 6$.

79. Since $x^2 = 18$, we obtain $x = \sqrt{18} = 3\sqrt{2}$. The base of a logarithm is positive.

81. Since $x + 1 = \log_3(7)$, we find
$$x = \log_3(7) - 1.$$

83. Since $y = \log(x)$ is one-to-one, we get
$$
\begin{aligned}
x &= 6 - x^2 \\
x^2 + x - 6 &= 0 \\
(x + 3)(x - 2) &= 0 \\
x &= -3, 2.
\end{aligned}
$$
But $\log(-3)$ is undefined, so the solution is $x = 2$.

85. $x = \log(25) \approx 1.3979$

87. Solving for x, we obtain
$$
\begin{aligned}
e^{2x} &= 3 \\
2x &= \ln(3) \\
x &= \frac{1}{2}\ln(3) \\
x &\approx 0.5493.
\end{aligned}
$$

89. Solving for x, we get
$$
\begin{aligned}
e^x &= \frac{4}{5} \\
x &= \ln\left(\frac{4}{5}\right) \\
x &\approx -0.2231.
\end{aligned}
$$

91 Solving for x, we find
$$
\begin{aligned}
\frac{1}{10^x} &= 2 \\
\frac{1}{2} &= 10^x \\
x &= \log\left(\frac{1}{2}\right) \\
x &\approx -0.3010.
\end{aligned}
$$

93. Solving for the year t when \$10 grows to \$20, we find
$$
\begin{aligned}
20 &= 10e^{rt} \\
2 &= e^{rt} \\
\ln 2 &= rt \\
\frac{\ln 2}{r} &= t.
\end{aligned}
$$

a) If $r = 2\%$, then $t = \dfrac{\ln 2}{0.02} \approx 34.7$ years.

b) If $r = 4\%$, then $t = \dfrac{\ln 2}{0.04} \approx 17.3$ years.

c) If $r = 8\%$, then $t = \dfrac{\ln 2}{0.08} \approx 8.7$ years.

d) If $r = 16\%$, then $t = \dfrac{\ln 2}{0.16} \approx 4.3$ years.

95. Solving for the annual percentage rate r when $10 becomes $30, we find

$$
\begin{align}
30 &= 10e^{rt} \\
3 &= e^{rt} \\
\ln 3 &= rt \\
\frac{\ln 3}{t} &= r.
\end{align}
$$

a) If $t = 5$ years, then $r = \dfrac{\ln 3}{5} \approx 0.2197 \approx 22\%$.

b) If $t = 10$ years, then $r = \dfrac{\ln 3}{10} \approx 0.10986 \approx 11\%$.

c) If $t = 20$ years, then $r = \dfrac{\ln 3}{20} \approx 0.0549 \approx 5.5\%$.

d) If $t = 40$ years, then $r = \dfrac{\ln 3}{40} \approx 0.027465 \approx 2.7\%$.

97. Let t be the number of years.

$$
\begin{align}
1000 \cdot e^{0.14t} &= 10^6 \\
e^{0.14t} &= 1000 \\
0.14t &= \ln(1000) \\
t &\approx 49.341 \text{ years}
\end{align}
$$

Note, $0.341(365) \approx 125$.
It will take 49 years and 125 days.

99. Since $e^{rt} = A/P$, $rt = \ln(A/P)$ and

$$
r = \frac{\ln(A/P)}{t}.
$$

Thus, $1000 will double in 3 years if the rate is $r = \dfrac{\ln(2000/1000)}{3} \approx 0.231$ or 23.1%.

101.

(a) $0.1e^{0.46(15)} \approx 99.2$ billion gigabits per year.

(b) Solving for t, we find

$$
\begin{align}
.1e^{.46t} &= 14 \\
e^{.46t} &= 140 \\
.46t &= \ln(140) \\
t &= \frac{\ln(140)}{.46} \\
t &\approx 11.
\end{align}
$$

In 2005 ($\approx 1994 + 11$), the data transmission is expected to be 14 billion gigabits/yr.

For Thought

1. False, since $\log(8) - \log(3) = \log(8/3) \neq \dfrac{\log(8)}{\log(3)}$.

2. True, since $\ln(3^{1/2}) = \dfrac{1}{2} \cdot \ln(3) = \dfrac{\ln(3)}{2}$.

3. True, since $\dfrac{\log_{19}(8)}{\log_{19}(2)} = \log_2(8) = 3 = \log_3(27)$.

4. True, because of the base-change formula.

5. False

6. False, since $\log(x) - \log(2) = \log(x/2)$.

7. False, since the solution of the first equation is $x = -2$ and the second equation is not defined when $x = -2$.

8. True   **9.**   False, since x can be negative.

10. False, since a can be negative and so $\ln(a)$ will not be a real number.

4.3 Exercises

1. $\sqrt{y}$ **3.** $y + 1$ **5.**   999

7. $\log(15)$

9. $\log_2((x - 1)x) = \log_2(x^2 - x)$

11. $\log_4(6)$ **13.** $\ln\left(\dfrac{x^8}{x^3}\right) = \ln(x^5)$

15. $\log_2(3x) = \log_2(3) + \log_2(x)$

17. $\log\left(\dfrac{x}{2}\right) = \log(x) - \log(2)$

19. $\log((x - 1)(x + 1)) = \log(x - 1) + \log(x + 1)$

21. $\log_a(5^3) = 3\log_a(5)$

23. $\log_a(5^{1/2}) = \dfrac{1}{2} \cdot \log_a(5)$

25. $\log_a(5^{-1}) = -\log_a(5)$

27. $\log_a(2) + \log_a(5)$

29. $\log_a(5/2) = \log_a(5) - \log_a(2)$

31. $\log_a(\sqrt{2^2 \cdot 5}) = \frac{1}{2}(2\log_a(2) + \log_a(5)) = \log_a(2) + \frac{1}{2}\log_a(5)$

33. $\log_3(5) + \log_3(x)$

35. $\log_2(5) - \log_2(2y) = \log_2(5) - \log_2(2) - \log_2(y)$

37. $\log(3) + \frac{1}{2}\log(x)$

39. $\log(3) + (x-1)\log(2)$

41. $\frac{1}{3} \cdot \ln(xy) - \frac{4}{3}\ln(t) = \frac{1}{3} \cdot \ln(x) + \frac{1}{3} \cdot \ln(y) - \frac{4}{3}\ln(t)$

43. $\log_2(5x^3)$

45. $\log_7(x^5) - \log_7(x^8) = \log(x^5/x^8) = \log_7(x^{-3})$

47. $\log(2xy/z)$

49. $\log\left(\frac{\sqrt{x}}{y}\right) + \log\left(\frac{z}{\sqrt[3]{w}}\right) = \log\left(\frac{z\sqrt{x}}{y\sqrt[3]{w}}\right)$

51. Since $2^x = 9$, we get $x = \dfrac{\log 9}{\log 2} \approx 3.1699$.

53. Since $0.56^x = 8$, we get
$$x = \frac{\log 8}{\log 0.56} \approx -3.5864.$$

55. Since $1.06^x = 2$, we get
$$x = \frac{\log 2}{\log 1.06} \approx 11.8957.$$

57. $\dfrac{\ln(9)}{\ln(4)} \approx \dfrac{2.1972246}{1.3862944} \approx 1.5850$

59. $\dfrac{\ln(2.3)}{\ln(9.1)} \approx \dfrac{0.8329091}{2.2082744} \approx 0.3772$

61. $\dfrac{\ln(12)}{\ln(1/2)} \approx -3.5850$

63. Since $4t = \log_{1.02}(3) = \dfrac{\ln(3)}{\ln(1.02)}$,

we find
$$t = \frac{\ln(3)}{4 \cdot \ln(1.02)} \approx 13.8695.$$

65. Since $365t = \log_{1.0001}(3.5) = \dfrac{\ln(3.5)}{\ln(1.0001)}$,

we get
$$t = \frac{\ln(3.5)}{365 \cdot \ln(1.0001)} \approx 34.3240.$$

67. Since $x^5 = 33.4$, we get $x = \sqrt[5]{33.4} \approx 2.0172$.

69. Since $x^{-1.3} = 0.546$, we have $x = 0.546^{1/(-1.3)}$ or $x \approx 1.5928$.

71. Let t be the number of years.
$$
\begin{aligned}
800\left(1 + \frac{0.08}{365}\right)^{365t} &= 2000 \\
\left(1 + \frac{0.08}{365}\right)^{365t} &= 2.5 \\
(1.0002192)^{365t} &\approx 2.5 \\
365t &\approx \log_{1.0002192}(2.5) \\
t &\approx \frac{1}{365} \cdot \frac{\ln(2.5)}{\ln(1.0002192)} \\
t &\approx 11.454889 \\
t &\approx 11 \text{ years, } 166 \text{ days}
\end{aligned}
$$

73. Let t be the number of years.
$$
\begin{aligned}
W\left(1 + \frac{0.1}{4}\right)^{4t} &= 3W \\
(1.025)^{4t} &= 3 \\
4t &\approx \log_{1.025}(3) \\
t &\approx \frac{1}{4} \cdot \frac{\ln(3)}{\ln(1.025)} \\
t &\approx 11.123 \text{ years} \\
t &\approx 11.123(4) \approx 44 \text{ quarters}
\end{aligned}
$$

75. Let t be the number of years.
$$
\begin{aligned}
4000(1 + r)^{200} &= 4.5 \times 10^6 \\
(1 + r)^{200} &= 1125 \\
r &= \sqrt[200]{1125} - 1 \\
r &\approx 0.035752 \text{ or } 3.58\%
\end{aligned}
$$

77. The Richter scale rating is
$$\log(I) - \log(I_o) = \log\left(\frac{I}{I_o}\right).$$

When $I = 1000 \cdot I_o$, the Richter scale rating is
$$\log\left(\frac{1000 \cdot I_o}{I_o}\right) = \log(1000) = 3.$$

79. $t = \dfrac{1}{r}\ln(P/P_o) = \dfrac{1}{r}\ln(P) - \dfrac{1}{r}\ln(P_o)$.

For Thought

1. True, since $(1.02)^x = 7$ is equivalent to $x = \log_{1.02}(7)$.

2. True, for $x(1 - \ln(3)) = 8$ implies $x = \dfrac{8}{1 - \ln(3)}$.

3. False, $\ln(1 - \sqrt{6})$ is undefined.

4. True, by the definition of a logarithm.

5. False, the exact solution is $x = \log_3(17)$ which is not the same as 2.5789.

6. False, since $x = -2$ is not a solution of the first equation but is a solution of the second one.

7. True, since $4^x = 2^{2x} = 2^{x-1}$ is equivalent to $2x = x - 1$.

8. True, since we may take the ln of both sides of $1.09^x = 2.3$.

9. True, since $\dfrac{\ln(2)}{\ln(7)} = \dfrac{\log(2)}{\log(7)}$.

10. True, since $\log(e) \cdot \ln(10) = \ln\left(10^{\log(e)}\right) = \ln(e) = 1$.

4.4 Exercises

1. Since $x = 2^3$, we get $x = 8$.

3. Since $10^2 = x + 20$, $x = 80$.

5. Since $10^1 = x^2 - 15$, we get $x^2 = 25$. The solutions are $x = \pm 5$.

7. Note, $x^2 = 9$ and $x = \pm 3$. Since the base of a logarithm is positive, the solution is $x = 3$.

9. Note, $x^{-2} = 4$ or $\dfrac{1}{4} = x^2$. Since the base of a logarithm is positive, the solution is $x = \dfrac{1}{2}$.

11. Since $x^3 = 10$, the solution is $x = \sqrt[3]{10}$.

13. Since $x = \left(8^{-1/3}\right)^2$, the solution is $x = \dfrac{1}{4}$.

15.

$$
\begin{aligned}
\log_2(x^2 - 4) &= 5 \\
x^2 - 4 &= 2^5 \\
x^2 &= 36 \\
x &= \pm 6
\end{aligned}
$$

Checking $x = -6$, one gets $\log_2(-6 + 2)$ which is undefined. The solution is $x = 6$.

17.

$$
\begin{aligned}
\log(5) + \log(x) &= 2 \\
\log(5x) &= 2 \\
5x &= 10^2 \\
x &= 20
\end{aligned}
$$

19. Since $\ln(x(x + 2)) = \ln(8)$ and $y = \ln(x)$ is one-to-one, we get

$$
\begin{aligned}
x^2 + 2x &= 8 \\
x^2 + 2x - 8 &= 0 \\
(x + 4)(x - 2) &= 0 \\
x &= -4, 2
\end{aligned}
$$

Since $\ln(-4)$ is undefined, the solution is $x = 2$.

21. Since $\log(4x) = \log\left(\dfrac{5}{x}\right)$ and $y = \log(x)$ is one-to-one, we obtain

$$
\begin{aligned}
4x &= \dfrac{5}{x} \\
4x^2 &= 5 \\
x^2 &= \dfrac{5}{4} \\
x &= \pm\dfrac{\sqrt{5}}{2}.
\end{aligned}
$$

But $\log\left(-\dfrac{\sqrt{5}}{2}\right)$ is undefined, so $x = \dfrac{\sqrt{5}}{2}$.

23. Since $\log_2\left(\dfrac{x}{3x - 1}\right) = 0$, we get

$$
\begin{aligned}
\dfrac{x}{3x - 1} &= 1 \\
x &= 3x - 1 \\
1 &= 2x \\
x &= \dfrac{1}{2}.
\end{aligned}
$$

25.

$$
\begin{aligned}
x \cdot \ln(3) + x \cdot \ln(2) &= 2 \\
x(\ln(3) + \ln(2)) &= 2 \\
&= \frac{2}{\ln(3) + \ln(2)} \\
x &= \frac{2}{\ln(6)}
\end{aligned}
$$

27. Since $x - 1 = \log_2(7)$, $x = \dfrac{\ln(7)}{\ln(2)} + 1 \approx 3.8074$.

29. Since $4x = \log_{1.09}(3.4)$, we find

$$
x = \frac{1}{4} \cdot \frac{\ln(3.4)}{\ln(1.09)} \approx 3.5502.
$$

31. Since $-x = \log_3(30)$, we obtain

$$
x = -\frac{\ln(30)}{\ln(3)} \approx -3.0959.
$$

33. Note, $-3x^2 = \ln(9)$. There is no solution since the left-hand side is non-negative and the right-hand side is positive.

35.

$$
\begin{aligned}
\ln(6^x) &= \ln(3^{x+1}) \\
x \cdot \ln(6) &= (x + 1) \cdot \ln(3) \\
x \cdot \ln(6) &= x \cdot \ln(3) + \ln(3) \\
x(\ln(6) - \ln(3)) &= \ln(3) \\
x &= \frac{\ln(3)}{\ln(6) - \ln(3)} \\
x &\approx 1.5850
\end{aligned}
$$

37.

$$
\begin{aligned}
\ln(e^{x+1}) &= \ln(10^x) \\
(x + 1) \cdot \ln(e) &= x \cdot \ln(10) \\
x + 1 &= x \cdot \ln(10) \\
1 &= x(\ln(10) - 1) \\
x &= \frac{1}{\ln(10) - 1} \\
x &\approx 0.7677
\end{aligned}
$$

39.

$$
\begin{aligned}
2^{x-1} &= (2^2)^{3x} \\
2^{x-1} &= 2^{6x} \\
x - 1 &= 6x \\
-1 &= 5x \\
x &= -0.2
\end{aligned}
$$

41.

$$
\begin{aligned}
\ln(6^{x+1}) &= \ln(12^x) \\
(x + 1) \cdot \ln(6) &= x \cdot \ln(12) \\
x \cdot \ln(6) + \ln(6) &= x \cdot \ln(12) \\
\ln(6) &= x(\ln(12) - \ln(6)) \\
x &= \frac{\ln(6)}{\ln(12) - \ln(6)} \\
x &\approx 2.5850
\end{aligned}
$$

43. Since $3 = e^{-\ln(w)} = e^{\ln(1/w)} = 1/w$, we have $\dfrac{1}{w} = 3$ and $w = \dfrac{1}{3}$.

45.

$$
\begin{aligned}
(\log(z))^2 &= 2 \cdot \log(z) \\
(\log(z))^2 - 2 \cdot \log(z) &= 0 \\
\log(z) \cdot (\log(z) - 2) &= 0 \\
\log(z) = 0 \quad &\text{or} \quad \log(z) = 2 \\
z = 10^0 \quad &\text{or} \quad z = 10^2 \\
z = 1 \quad &\text{or} \quad z = 100
\end{aligned}
$$

The solutions are $z = 1, 100$.

47. Divide the equation by $4(1.03)^x$.

$$
\begin{aligned}
\left(\frac{1.02}{1.03}\right)^x &= \frac{3}{4} \\
\ln\left(\left(\frac{1.02}{1.03}\right)^x\right) &= \ln\left(\frac{3}{4}\right) \\
x \cdot \ln\left(\frac{1.02}{1.03}\right) &= \ln\left(\frac{3}{4}\right) \\
x &= \frac{\ln\left(\frac{3}{4}\right)}{\ln\left(\frac{1.02}{1.03}\right)} \\
x &\approx 29.4872
\end{aligned}
$$

49. Note that $e^{\ln((x^2)^3)-\ln(x^2)} = e^{\ln(x^6)-\ln(x^2)} = e^{\ln(x^6/x^2)} = e^{\ln(x^4)} = x^4$.

Thus, $x^4 = 16$ and $x = \pm 2$.

But $\ln(-2)$ is undefined, so $x = 2$.

51. Since $\left(\dfrac{1}{2}\right)^2 = \dfrac{1}{4}$, we find

$$
\begin{aligned}
\left(\frac{1}{2}\right)^{2x-1} &= \left(\frac{1}{2}\right)^{6x+4} \\
2x - 1 &= 6x + 4 \\
-5 &= 4x \\
x &= -\frac{5}{4}.
\end{aligned}
$$

53. By approximating the x-intercepts of the graph $y = 2^x - 3^{x-1} - 5^{-x}$, we find that the solutions are $x \approx 0.194, 2.70$.

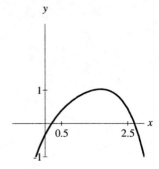

55. By approximating the x-intercept of the graph $y = \ln(x + 51) - \log(-48 - x)$, we obtain that the solution is $x \approx -49.73$.

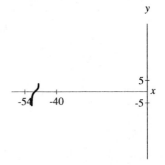

57. By approximating the x-intercepts of the graph $y = x^2 - 2^x$, we find that the solutions are $x \approx -0.767, 2, 4$.

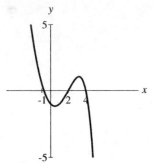

59. Solving for the rate of decay r, we find

$$
\begin{aligned}
A_0/2 &= A_0 e^{10,000r} \\
1/2 &= e^{10,000r} \\
\ln(1/2) &= 10,000r \\
\frac{\ln(1/2)}{10,000} &= r
\end{aligned}
$$

Approximately, $r \approx -6.93 \times 10^{-5}$.

61. Using $A = A_o e^{rt}$ with $A_o = 1$ and the half-life, we obtain $\dfrac{1}{2} = e^{5730r}$. Thus, $5730r = \ln\left(\dfrac{1}{2}\right)$ and $r \approx -0.000120968$. When $A = 0.1$, we get

$$
\begin{aligned}
0.1 &= e^{-0.000120968t} \\
\ln(0.1) &= -0.000120968t \\
t = \frac{\ln(0.1)}{-0.000120968} &\approx 19,035 \text{ years}
\end{aligned}
$$

63. From Number 61, $r \approx -0.000120968$. When $A = 10$ and $A_o = 12$, we obtain

$$
\begin{aligned}
10 &= 12 \cdot e^{-0.000120968t} \\
\ln(10/12) &= -0.000120968t \\
t = \frac{\ln(5/6)}{-0.000120968} &\approx 1507 \text{ years}.
\end{aligned}
$$

65. Let $A = A_o e^{rt}$ where $A_o = 25$, $A = 20$, and $t = 8000$. Then

$$
\begin{aligned}
20 &= 25 \cdot e^{8000r} \\
\ln(0.8) &= 8000r \\
r &\approx -0.000027893.
\end{aligned}
$$

To find the half-life, let $A = 12.5$. Thus, we have

$$12.5 = 25 \cdot e^{-0.000027893t}$$
$$\ln(0.5) = -0.000027893t$$
$$t \approx 24,850 \text{ years.}$$

67. $\dfrac{2.5(0.5)^{24/14}}{2.5} \times 100 \approx 30.5\%$, the percentage of the last dosage that remains before the next dosage is taken

69. The initial difference in temperature is $325 - 35 = 290$, and after $t = 3$ hours the difference is $325 - 140 = 185$. Then

$$185 = 290 \cdot e^{3k}$$
$$\ln\left(\frac{185}{290}\right) = 3k$$
$$k \approx -0.1498417.$$

The difference in temperature when the roast is well-done is $325 - 170 = 155$. Thus,

$$155 = 290 \cdot e^{(-0.1498417)\cdot t}$$
$$\ln\left(\frac{155}{290}\right) = (-0.1498417) \cdot t$$
$$t \approx 4.18 \text{ hr}$$
$$t \approx 4 \text{ hr and } 11 \text{ min.}$$

James must wait 1 hour and 11 minutes longer.

If the oven temperature is set at 170°, then the initial and final differences are 135 and 0, respectively. Since $0 = 135 \cdot e^{(-0.1498417)\cdot t}$ has no solution, James has to wait forever.

71. At 7:00 a.m., the difference in temperature is $80 - 40 = 40$, and $t = 1$ hour later the difference in temperature is $72 - 40 = 32$. Then

$$32 = 40 \cdot e^{1\cdot k}$$
$$\ln\left(\frac{32}{40}\right) = k$$
$$k \approx -0.2231436.$$

Let n be the number of hours before 7:00 a.m. when death occured. At the time of death, the difference in temperature is $98 - 40 = 58$.

Then

$$40 = 58 \cdot e^{-0.2231436 \cdot n}$$
$$\ln\left(\frac{40}{58}\right) = -0.2231436 \cdot n$$
$$n \approx 1.665 \text{ hr}$$
$$n \approx 1 \text{ hr and } 40 \text{ min.}$$

Death occured at 5:20 a.m.

73. Since $R = P\dfrac{i}{1 - (1+i)^{-nt}}$, we obtain

$$1 - (1+i)^{-nt} = Pi/R$$
$$1 - Pi/R = (1+i)^{-nt}$$
$$\ln(1 - Pi/R) = -nt\ln(1+i)$$
$$\frac{-\ln(1 - Pi/R)}{n\ln(1+i)} = t.$$

Let $i = 0.09/12 = 0.0075$, $P = 100,000$, and $R = 1250$. Then

$$t = \frac{-\ln(1 - Pi/R)}{n\ln(1+i)} \approx 10.219.$$

It will take 10 yr, 3 mo to pay off the loan.

75. The future values of the $1000 and $1100 investments are equal. Then

$$1,000 \cdot e^{0.06t} = 1100\left(1 + \frac{0.06}{365}\right)^{365t}$$
$$e^{0.06t} = 1.1\left(1 + \frac{0.06}{365}\right)^{365t}$$
$$.06t = \ln(1.1) + 365t\ln\left(1 + \frac{0.06}{365}\right)$$

$$.06t - 365t\ln\left(1 + \frac{0.06}{365}\right) = \ln(1.1)$$

$$t = \frac{\ln(1.1)}{.06 - 365\ln\left(1 + \frac{0.06}{365}\right)}$$
$$t \approx 19,328.84173 \text{ years}$$

They will be equal after $19,328$ yr, 307 days.

77. a) Let $n = 2500$ and $A = 400$. Since $n = k \log(A)$, we find

$$2500 = k \log(400).$$

Then $k = \dfrac{2500}{\log(400)}$. When $A = 200$, the number of species left is

$$n = \frac{2500}{\log(400)} \log(200) \approx 2211 \text{ species.}$$

b) Let $n = 3500$ and $A = 1200$. Since $n = k \log(A)$, we obtain

$$3500 = k \log(1200).$$

Then $k = \dfrac{3500}{\log(1200)}$. When $n = 1000$, the remaining forest area is given by

$$1000 = \frac{3500}{\log(1200)} \log(A)$$

$$\frac{1000}{3500} \log(1200) = \log(A)$$

$$\frac{2}{7} \log(1200) = \log(A)$$

$$\log\left(1200^{2/7}\right) = \log(A)$$

$$1200^{2/7} = A$$

Thus, the percentage of forest that has been destroyed is

$$100 - \frac{A}{1200}(100) \approx 99\%.$$

79. If the sound level is 90 db, then the intensity of the sound is given by

$$
\begin{aligned}
10 \cdot \log(I \times 10^{12}) &= 90 \\
\log(I) + \log(10^{12}) &= 9 \\
\log(I) + 12 &= 9 \\
\log(I) &= -3 \\
I &= 10^{-3} \text{ watts/m}^2.
\end{aligned}
$$

81. Since $P \cdot e^{(0.06)18} = 20,000$, the investment will grow to $P = \dfrac{20,000}{e^{(0.06)18}} \approx \$6791.91$.

Chapter 4 Review Exercises

1. 64   **3.** 6   **5.** 0

7. 17   **9.** $2^5 = 32$ **11.** $\log(10^3) = 3$

12. $10^{\log(5)} = 5$ **13.** $\log_2(2^9) = 9$

15. $\log(1000) = 3$

17. $\log_2(1) - \log_2(8) = 0 - 3 = -3$

19. $\log_2(8) = 3$

21. $\log((x-3)x) = \log(x^2 - 3x)$

23. $\ln(x^2) + \ln(3y) = \ln(3x^2 y)$

25. $\log(3) + \log(x^4) = \log(3) + 4 \cdot \log(x)$

27. $\log_3(5) + \log_3(x^{1/2}) - \log_3(y^4) =$
     $\log_3(5) + \dfrac{1}{2} \cdot \log_3(x) - 4 \cdot \log_3(y)$

29. $\ln(2 \cdot 5) = \ln(2) + \ln(5)$

31. $\ln(5^2 \cdot 2) = \ln(5^2) + \ln(2) = 2 \cdot \ln(5) + \ln(2)$

33. Since $\log_{10}(x) = 10$, we get $x = 10^{10}$.

35. Since $x^4 = 81$ and $x > 0$, $x = 3$.

37. Since $\log_{1/3}(27) = -3 = x + 2$, we get $x = -5$.

39. Since $3^{x+2} = 3^{-2}$, $x + 2 = -2$. So $x = -4$.

41. Since $x - 2 = \ln(9)$, we obtain $x = 2 + \ln(9)$.

43. Since $(2^2)^{x+3} = 2^{2x+6} = 2^{-x}$, we get $2x + 6 = -x$. Then $6 = -3x$ and so $x = -2$.

45.

$$
\begin{aligned}
\log(2x^2) &= 5 \\
2x^2 &= 10^5 \\
x^2 &= 50,000 \\
x &= \pm 100\sqrt{5}
\end{aligned}
$$

Since $\log(-100\sqrt{5})$ is undefined, $x = 100\sqrt{5}$.

47.

$$
\begin{aligned}
\log_2\left(x^2 - 4x\right) &= \log_2(x + 24) \\
x^2 - 4x &= x + 24 \\
x^2 - 5x - 24 &= 0 \\
(x - 8)(x + 3) &= 0 \\
x &= 8, -3
\end{aligned}
$$

Since $\log(-3)$ is undefined, $x = 8$.

49. Since $\ln((x+2)^2) = \ln(4^3)$ and $y = \ln(x)$ is a one-to-one function, we obtain

$$
\begin{aligned}
(x+2)^2 &= 64 \\
x+2 &= \pm 8 \\
x &= -2 \pm 8 \\
x &= 6, -10.
\end{aligned}
$$

Checking $x = -10$ one gets $2\ln(-8)$ which is undefined. So $x = 6$.

51.

$$
\begin{aligned}
x \cdot \log(4) + x \cdot \log(25) &= 6 \\
x(\log(4) + \log(25)) &= 6 \\
x \cdot \log(100) &= 6 \\
x \cdot 2 &= 6 \\
x &= 3
\end{aligned}
$$

53. The missing coordinates are

(i) 3 since $\left(\dfrac{1}{3}\right)^{-1} = 3$,

(ii) -3 since $\left(\dfrac{1}{3}\right)^{-3} = 27$,

(iii) $\sqrt{3}$ since $\left(\dfrac{1}{3}\right)^{-1/2} = \sqrt{3}$, and

(iv) 0 since $\left(\dfrac{1}{3}\right)^{0} = 1$.

55. c

57. b

59. d

61. e

63. Domain $(-\infty, \infty)$, range $(0, \infty)$, increasing, asymptote $y = 0$

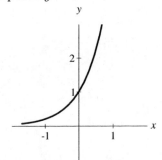

65. Domain $(-\infty, \infty)$, range $(0, \infty)$, decreasing, asymptote $y = 0$

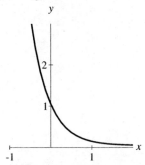

67. Domain $(0, \infty)$, range $(-\infty, \infty)$, increasing, asymptote $x = 0$

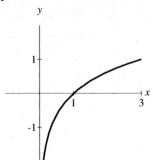

69. Domain $(-3, \infty)$, range $(-\infty, \infty)$, increasing, asymptote $x = -3$

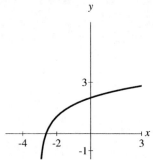

71. Domain $(-\infty, \infty)$, range $(1, \infty)$, increasing, asymptote $y = 1$

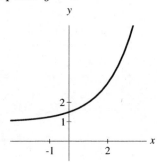

73. Domain $(-\infty, 2)$, range $(-\infty, \infty)$, decreasing, asymptote $x = 2$

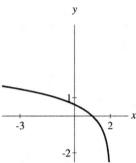

75. $f^{-1}(x) = \log_7(x)$

77. $f^{-1}(x) = 5^x$

79. Replace $f(x)$ by y, interchange x and y, solve for y, and replace y by $f^{-1}(x)$.

$$
\begin{aligned}
y &= 3\log(x-1) \\
x &= 3\log(y-1) \\
\frac{x}{3} &= \log(y-1) \\
10^{x/3} &= y-1 \\
10^{x/3}+1 &= y \\
f^{-1}(x) &= 10^{x/3}+1
\end{aligned}
$$

81. Replace $f(x)$ by y, interchange x and y, solve for y, and replace y by $f^{-1}(x)$.

$$
\begin{aligned}
y &= e^{x+2} - 3 \\
x &= e^{y+2} - 3 \\
x + 3 &= e^{y+2} \\
\ln(x+3) &= y + 2 \\
\ln(x+3) - 2 &= y \\
f^{-1}(x) &= \ln(x+3) - 2
\end{aligned}
$$

83. Since $3^x = 10$, $x = \log_3(10) = \dfrac{\ln(10)}{\ln(3)} \approx 2.0959$

.

85. Since $\log_3(x) = 1.876$, $x = 3^{1.876} \approx 7.8538$.

87. After taking the natural logarithm of both sides, we have

$$
\begin{aligned}
\ln(5^x) &= \ln(8^{x+1}) \\
x \cdot \ln(5) &= (x+1) \cdot \ln(8)
\end{aligned}
$$

$$
\begin{aligned}
x \cdot \ln(5) &= x \cdot \ln(8) + \ln(8) \\
x \cdot (\ln(5) - \ln(8)) &= \ln(8) \\
x &= \frac{\ln(8)}{\ln(5) - \ln(8)} \\
x &\approx -4.4243.
\end{aligned}
$$

89. True, since $\log_3(81) = 4$ and $2 = \log_3(9)$.

91. False, since $\ln(3^2) = 2 \cdot \ln(3) \neq (\ln(3))^2$.

93. True, since $4 \cdot \log_2(8) = 4 \cdot 3 = 12$.

95. False, since $3 + \log(6) = \log(10^3) + \log(6) = \log(6000)$.

97. False, since $\log_2(16) = 4$, $\log_2(8) = 3$, and $\dfrac{3}{4} \neq 3 - 4$.

99. False, since $\log_2(25) = 2\log_2(5) = 2 \cdot \dfrac{\log(5)}{\log(2)} \neq 2 \cdot \log(5)$.

101. True, because of the base-changing formula.

103. Solving for r, we find:

$$
\begin{aligned}
\frac{A}{P} &= e^{rt} \\
\ln\frac{A}{P} &= rt \\
\frac{\ln(A/P)}{t} &= r.
\end{aligned}
$$

105. The value at the end of 18 years is

$$
50,000\left(1 + \frac{0.05}{4}\right)^{18\cdot 4} \approx \$122,296.01.
$$

107. Let t be the number of years.

$$
\begin{aligned}
50,000\left(1 + \frac{0.05}{4}\right)^{4t} &= 100,000 \\
(1.0125)^{4t} &= 2 \\
4t &= \log_{1.0125}(2) \\
t &= \frac{1}{4} \cdot \frac{\ln(2)}{\ln(1.0125)} \\
t &\approx 13.9 \ \text{years}
\end{aligned}
$$

It doubles in $4 \cdot 13.9 \approx 56$ quarters.

109. The present amount is $A = 25 \cdot e^0 = 25$ g.
After $t = 1000$ years, the amount left
is $A = 25 \cdot e^{-0.32} \approx 18.15$ g.

To find the half-life, let $A = 12.5$.

$$
\begin{aligned}
25 \cdot e^{-0.00032t} &= 12.5 \\
e^{-0.00032t} &= 0.5 \\
-0.00032t &= \ln(0.5) \\
t &\approx 2166
\end{aligned}
$$

The half-life is 2166 years.

111. Let $f(t) = 10,000$.

$$
\begin{aligned}
40,000 \cdot (1 - e^{-0.0001t}) &= 10,000 \\
1 - e^{-0.0001t} &= 0.25 \\
0.75 &= e^{-0.0001t} \\
\ln(0.75) &= -0.0001t \\
t &\approx 2877 \text{ hr}
\end{aligned}
$$

It takes 2877 hours to learn 10,000 words.

113. In the following equations, we solve for x.

$$
\begin{aligned}
1026 \left(\frac{25,005}{64} \right)^x &= 19.2 \\
\left(\frac{25,005}{64} \right)^x &= \frac{19.2}{1026} \\
x \ln \left(\frac{25,005}{64} \right) &= \ln \left(\frac{19.2}{1026} \right) \\
x &= \frac{\ln \left(\frac{19.2}{1026} \right)}{\ln \left(\frac{25,005}{64} \right)} \\
x \approx -0.667 \quad \text{or} \quad x &\approx -\frac{2}{3}
\end{aligned}
$$

In the following equations, we obtain y.

$$
\begin{aligned}
\frac{25005}{2240} \left(\frac{35 + \frac{1}{12}}{100} \right)^y &= 258.51 \\
\left(\frac{35 + \frac{1}{12}}{100} \right)^y &= \frac{258.51(2240)}{25005} \\
y &= \frac{\ln \left(\frac{258.51(2240)}{25005} \right)}{\ln \left(\frac{35 + \frac{1}{12}}{100} \right)} \\
y &\approx -3
\end{aligned}
$$

Next, we solve for z.

$$
\begin{aligned}
13.5 \left(\frac{25005}{64} \right)^z &= 1.85 \\
\left(\frac{25005}{64} \right)^z &= \frac{1.85}{13.5} \\
z \ln \left(\frac{25005}{64} \right) &= \ln \left(\frac{1.85}{13.5} \right) \\
z &= \frac{\ln \left(\frac{1.85}{13.5} \right)}{\ln \left(\frac{25005}{64} \right)} \\
z \approx -0.333 \quad \text{or} \quad z &\approx -\frac{1}{3}
\end{aligned}
$$

For Thought

1. False, vertex is $(0, -1/2)$.   **2.**   True

3. True, since $p = 3/2$ and the focus is $(4, 5)$, vertex is $(4 - 3/2, 5) = (5/2, 5)$.

4. False, focus is at $(0, 1/4)$ since parabola opens upward.   **5.**   True, $p = 1/4$.

6. False. Since $p = 4$ and the vertex is $(2, -1)$,

equation of parabola is $y = \dfrac{1}{16}(x - 2)^2 - 1$

and x-intercepts are $(6, 0)$, $(-2, 0)$.

7. False, if $x = 0$ then $y = 0$ and y-intercept is $(0, 0)$.   **8.**   True

9. False, since $p = 1/4$, vertex is $(5, 4)$, and focus is $(5, 4 + 1/4) = (-5, 17/4)$.

10. False, it opens to the left.

5.1 Exercises

1. Note $p = 1$.

Vertex $(0, 0)$, focus $(0, 1)$, and directrix $y = -1$

3. Note $p = -1/2$. Vertex $(1, 2)$, focus $(1, 3/2)$, and directrix $y = 5/2$

5. Note $p = 3/4$. Vertex $(3, 1)$, focus $(15/4, 1)$, and directrix $x = 9/4$

7. Since the vertex is equidistant from $y = -2$ and $(0, 2)$, the vertex is $(0, 0)$ and $p = 2$. Then

$a = \dfrac{1}{4p} = \dfrac{1}{8}$ and an equation is $y = \dfrac{1}{8}x^2$.

9. Since the vertex is equidistant from $y = 3$ and $(0, -3)$, the vertex is $(0, 0)$ and $p = -3$. Then

$a = \dfrac{1}{4p} = -\dfrac{1}{12}$ and an equation is $y = -\dfrac{1}{12}x^2$.

11. One finds $p = \dfrac{3}{2}$. So $a = \dfrac{1}{4p} = \dfrac{1}{6}$ and

vertex is $\left(3, 5 - \dfrac{3}{2}\right) = \left(3, \dfrac{7}{2}\right)$. Parabola

is given by $y = \dfrac{1}{6}(x - 3)^2 + \dfrac{7}{2}$.

13. One finds $p = -\dfrac{5}{2}$. So $a = \dfrac{1}{4p} = -\dfrac{1}{10}$ and

vertex is $\left(1, -3 + \dfrac{5}{2}\right) = \left(1, -\dfrac{1}{2}\right)$. Parabola

is given by $y = -\dfrac{1}{10}(x - 1)^2 - \dfrac{1}{2}$.

15. One finds $p = 0.2$. So $a = \dfrac{1}{4p} = 1.25$ and

vertex is $(-2, 1.2 - 0.2) = (-2, 1)$. Parabola

is given by $y = 1.25(x + 2)^2 + 1$.

17. Since $p = 1$, we get $a = \dfrac{1}{4p} = \dfrac{1}{4}$.

An equation is $y = \dfrac{1}{4}x^2$.

19. Since $p = -\dfrac{1}{4}$, we get $a = \dfrac{1}{4p} = -1$.

An equation is $y = -x^2$

21. Note, vertex is $(1, 0)$ and since $a = \dfrac{1}{4p} = 1$,

we find $p = \dfrac{1}{4}$.   The focus is $(1, 0 + p) = \left(1, \dfrac{1}{4}\right)$ and the directrix is $y = 0 - p$ or

$y = -\dfrac{1}{4}$.

23. Note, vertex is $(3, 0)$ and since $a = \dfrac{1}{4p} = \dfrac{1}{4}$,

we find $p = 1$. The focus is $(3, 0 + p) = (3, 1)$

and the directrix is $y = 0 - p$ or $y = -1$.

25. Note, vertex is $(3, 4)$ and since $a = \dfrac{1}{4p} = -2$,

we find $p = -\dfrac{1}{8}$.   The focus is $(3, 4 + p) = \left(3, \dfrac{31}{8}\right)$ and the directrix is $y = 4 - p$ or

$y = \dfrac{33}{8}$.

27. Completing the square, we obtain

$$\begin{aligned} y &= (x^2 - 8x + 16) - 16 + 3 \\ y &= (x - 4)^2 - 13. \end{aligned}$$

Since $\dfrac{1}{4p} = 1$, $p = 0.25$.

Since vertex is $(4, -13)$, focus is $(4, -13 + 0.25) = (4, -51/4)$, and directrix is $y = -13 - p = -53/4$.

29. Completing the square, we obtain

$$
\begin{aligned}
y &= 2(x^2 + 6x + 9) + 5 - 18 \\
y &= 2(x + 3)^2 - 13.
\end{aligned}
$$

Since $\dfrac{1}{4p} = 2$, $p = 1/8$.

Since vertex is $(-3, -13)$, the focus is $(-3, -13 + 1/8) = (-3, -103/8)$, and directrix is $y = -13 - 1/8 = -105/8$.

31. Completing the square, we get

$$
\begin{aligned}
y &= -2(x^2 - 3x + 9/4) + 1 + 9/2 \\
y &= -2(x - 3/2)^2 + 11/2.
\end{aligned}
$$

Since $\dfrac{1}{4p} = -2$, $p = -1/8$.

Since vertex is $(3/2, 11/2)$, the focus is $(3/2, 11/2 - 1/8) = (3/2, 43/8)$, and directrix is $y = 11/2 + 1/8 = 45/8$.

33. Completing the square,

$$
\begin{aligned}
y &= 5(x^2 + 6x + 9) - 45 \\
y &= 5(x + 3)^2 - 45.
\end{aligned}
$$

Since $\dfrac{1}{4p} = 5$, we have $p = 0.05$.

Since vertex is $(-3, -45)$, the focus is $(-3, -45 + 0.05) = (-3, -44.95)$, and directrix is $y = -45 - .05 = -45.05$.

35. Completing the square, we get

$$
\begin{aligned}
y &= \frac{1}{8}(x^2 - 4x + 4) + \frac{9}{2} - \frac{1}{2} \\
y &= \frac{1}{8}(x - 2)^2 + 4.
\end{aligned}
$$

Since $\dfrac{1}{4p} = 1/8$, we find $p = 2$.

Since vertex is $(2, 4)$, the focus is $(2, 4 + 2) = (2, 6)$, and directrix is $y = 4 - 2$ or $y = 2$.

37. Since $a = 1$ and $b = -4$, we find $x = \dfrac{-b}{2a} = \dfrac{4}{2} = 2$. Since $\dfrac{1}{4p} = a = 1$, we obtain $p = 1/4$. Substituting $x = 2$, we get $y = 2^2 - 4(2) + 3 = -1$. Thus, the vertex is $(2, -1)$, the focus is $(2, -1 + p) = (2, -3/4)$, the directrix is $y = -1 - p = -5/4$, and the parabola opens upward since $a > 0$.

39. Note, $a = -1$, $b = 2$. So $x = \dfrac{-b}{2a} = \dfrac{-2}{-2} = 1$ and since $\dfrac{1}{4p} = a = -1$, $p = -1/4$.

Substituting $x = 1$, $y = -(1)^2 + 2(1) - 5 = -4$. The vertex is $(1, -4)$ and focus is $(1, -4 + p) = (1, -17/4)$, directrix is $y = -4 - p = -15/4$, and parabola opens down since $a < 0$.

41. Note, $a = 3$, $b = -6$. So $x = \dfrac{-b}{2a} = \dfrac{6}{6} = 1$ and since $\dfrac{1}{4p} = a = 3$, $p = 1/12$. Substituting $x = 1$, $y = 3(1)^2 - 6(1) + 1 = -2$. The vertex is $(1, -2)$ and focus is $(1, -2 + p) = (1, -23/12)$, directrix is $y = -2 - p = -25/12$, and parabola opens up since $a > 0$.

43. Note, $a = -1/2$, $b = -3$. So $x = \dfrac{-b}{2a} = \dfrac{3}{-1} = -3$ and since $\dfrac{1}{4p} = a = -1/2$, $p = -1/2$. Substituting $x = -3$, we have $y = -\dfrac{1}{2}(-3)^2 - 3(-3) + 2 = \dfrac{13}{2}$. The vertex is $\left(-3, \dfrac{13}{2}\right)$ and focus is $\left(-3, \dfrac{13}{2} + p\right) = (-3, 6)$, directrix is $y = \dfrac{13}{2} - p = 7$, and parabola opens down since $a < 0$.

45. Note, $y = \frac{1}{4}x^2 + 5$ is of the form

$y = a(x-h)^2 + k$. So $h = 0$, $k = 5$, and

$\frac{1}{4p} = a = \frac{1}{4}$ from which we have $p = 1$. The

vertex is $(h,k) = (0,5)$, focus is $(0, 5+p) = (0,6)$, directrix is $y = 5 - p = 4$ and parabola opens up since $a > 0$.

47. From the given focus and directrix, one

finds $p = 1/4$. So $a = \frac{1}{4p} = 1$, vertex is

$(h,k) = (1/2, -2-p) = (1/2, -9/4)$,

axis of symmetry is $x = 1/2$, and parabola

is given by $y = a(x-h)^2 + k = \left(x - \frac{1}{2}\right)^2 - \frac{9}{4}$.

If $y = 0$ then $x - \frac{1}{2} = \pm\frac{3}{2}$ or $x = 2, -1$.

The x-intercepts are $(2,0), (-1,0)$.

If $x = 0$ then $y = \left(0 - \frac{1}{2}\right)^2 - \frac{9}{4} = -2$ and

y-intercept is $(0, -2)$.

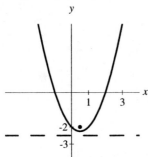

49. From the given focus and directrix one

finds $p = -1/4$. So $a = \frac{1}{4p} = -1$, vertex is

$(h,k) = (-1/2, 6-p) = (-1/2, 25/4)$,

axis of symmetry is $x = -1/2$, and parabola is given by

$$y = a(x-h)^2 + k = -\left(x + \frac{1}{2}\right)^2 + \frac{25}{4}.$$

If $y = 0$ then $x + \frac{1}{2} = \pm\frac{5}{2}$ or $x = -3, 2$.

The x-intercepts are $(-3, 0), (2, 0)$.

If $x = 0$ then $y = -\left(0 + \frac{1}{2}\right)^2 + \frac{25}{4} = 6$

and y-intercept is $(0, 6)$.

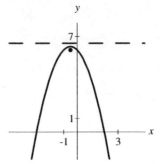

51. Since $\frac{1}{2}(x+2)^2 + 2$ is of the form $a(x-h)^2 + k$,

vertex is $(h,k) = (-2, 2)$ and axis of symmetry

is $x = -2$. If $y = 0$ then $0 = \frac{1}{2}(x+2)^2 + 2$; this

has no solution since left-hand side is always positive. No x-intercept. If $x = 0$ then

$y = \frac{1}{2}(0+2)^2 + 2 = 4$. y-intercept is $(0, 4)$.

Since $\frac{1}{4p} = a = \frac{1}{2}$, $p = \frac{1}{2}$, focus is $(h, k+p) = (-2, 5/2)$, and directrix is $y = k - p = 3/2$.

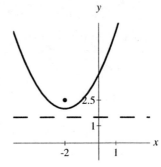

53. Since $-\frac{1}{4}(x+4)^2 + 2$ is of the form $a(x-h)^2 + k$,

vertex is $(h,k) = (-4, 2)$ and axis of symmetry is $x = -4$. If $y = 0$ then

$$\frac{1}{4}(x+4)^2 = 2$$
$$x + 4 = \pm\sqrt{8}.$$

x-intercepts are $(-4 \pm 2\sqrt{2}, 0)$. If $x = 0$, then

$y = -\frac{1}{4}(0+4)^2 + 2 = -2$. The y-intercept is

$(0, -2)$. Since $\frac{1}{4p} = a = -\frac{1}{4}$, $p = -1$,

focus is $(h, k + p) = (-4, 1)$, and directrix is $y = k - p = 3$.

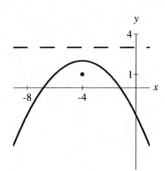

55. Since $\dfrac{1}{2}x^2 - 2$ is of the form $a(x-h)^2 + k$,

vertex is $(h, k) = (0, -2)$ and axis of symmetry is $x = 0$. If $y = 0$, then

$$\begin{aligned} \frac{1}{2}x^2 &= 2 \\ x^2 &= 4. \end{aligned}$$

x-intercepts are $(\pm 2, 0)$. If $x = 0$ then

$y = \dfrac{1}{2}(0)^2 - 2 = -2$. The y-intercept is $(0, -2)$.

Since $\dfrac{1}{4p} = a = \dfrac{1}{2}$, $p = 1/2$, focus is

$(h, k + p) = (0, -3/2)$, and directrix is $y = k - p = -5/2$.

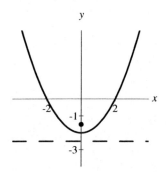

57. Since $y = (x-2)^2$ is of the form $a(x-h)^2 + k$, vertex is $(h, k) = (2, 0)$, and axis of symmetry is $x = 2$. If $y = 0$ then $(x-2)^2 = 0$ and x-intercept is $(2, 0)$. If $x = 0$ then $y = (0-2)^2 = 4$ and y-intercept is $(0, 4)$. Since $\dfrac{1}{4p} = a = 1$,

$p = 1/4$, focus is $(h, k + p) = (2, 1/4)$, and directrix is $y = k - p = -1/4$.

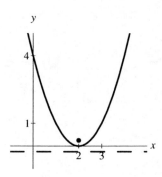

59. By completing the square, we obtain

$$y = \frac{1}{3}(x - 3/2)^2 - 3/4.$$

Vertex is $(h, k) = (3/2, -3/4)$ and axis of symmetry is $x = 3/2$. If $y = 0$, then

$$\begin{aligned} \frac{1}{3}\left(x - \frac{3}{2}\right)^2 &= \frac{3}{4} \\ \left(x - \frac{3}{2}\right)^2 &= \frac{9}{4} \\ x &= \frac{3}{2} \pm \frac{3}{2}. \end{aligned}$$

x-intercepts are $(3, 0), (0, 0)$. If $x = 0$, then

$y = \dfrac{1}{3}\left(0 - \dfrac{3}{2}\right)^2 - \dfrac{3}{4} = 0$ and y-intercept is

$(0, 0)$. Since $\dfrac{1}{4p} = a = 1/3$, $p = 3/4$, focus

is $(h, k + p) = (3/2, 0)$, and directrix is $y = k - p = -3/2$ or $y = -3/2$.

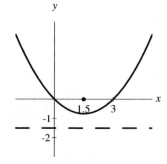

61. Since $x = -y^2$ is of the form $x = a(y-h)^2 + k$, vertex is $(k, h) = (0, 0)$ and axis of symmetry is $y = 0$. If $y = 0$ then $x = -0^2 = 0$ and x-intercept is $(0, 0)$. If $x = 0$ then $0 = -y^2$ and y-intercept is $(0, 0)$. Since $\dfrac{1}{4p} = a = -1$,

$p = -1/4$, focus is $(k + p, h) = (-1/4, 0)$, and directrix is $x = k - p = 1/4$.

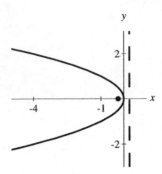

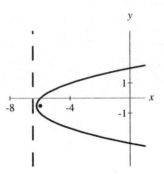

63. Since $x = -\dfrac{1}{4}y^2 + 1$ is of the form

$$x = a(y - h)^2 + k,$$

vertex is $(k, h) = (1, 0)$ and axis of symmetry is $y = 0$. If $y = 0$ then $x = 1$ and x-intercept is $(1, 0)$. If $x = 0$ then $\dfrac{1}{4}y^2 = 1$, $y^2 = 4$, and y-intercepts are $(0, \pm 2)$. Since $\dfrac{1}{4p} = a = -\dfrac{1}{4}$, $p = -1$, focus is $(k + p, h) = (0, 0)$, and directrix is $x = k - p = 2$ or $x = 2$.

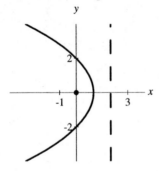

65. By completing the square, we obtain

$$x = (y + 1/2)^2 - 25/4.$$

Vertex is $(k, h) = (-25/4, -1/2)$ and axis of symmetry is $y = -1/2$. If $y = 0$, then $x = 1/4 - 25/4 = -6$. The x-intercept is $(-6, 0)$. If $x = 0$, then

$$\left(y + \dfrac{1}{2}\right)^2 = \dfrac{25}{4}$$
$$y = -\dfrac{1}{2} \pm \dfrac{5}{2}$$
$$y = 2, -3.$$

y-intercepts are $(0, 2), (0, -3)$. Since $\dfrac{1}{4p} = a = 1$, $p = 1/4$, focus is $(k + p, h) = (-6, -1/2)$, and directrix is $x = k - p = -13/2$.

67. By completing the square, we get

$$x = -\dfrac{1}{2}(y + 1)^2 - 7/2.$$

Vertex is $(k, h) = (-7/2, -1)$ and axis of symmetry is $y = -1$. If $y = 0$, then $x = -1/2 - 7/2 = -4$. The x-intercept is $(-4, 0)$. If $x = 0$, then

$$0 = -\dfrac{1}{2}(y + 1)^2 - 7/2 < 0$$

which is inconsistent and so there is no y-intercept. Since $\dfrac{1}{4p} = a = -1/2$, $p = -1/2$, focus is $(k + p, h) = (-4, -1)$, and directrix is $x = k - p = -3$.

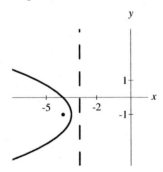

69. Since $x = 2(y - 1)^2 + 3$ is of the form $a(y - h)^2 + k$, we find that the vertex is $(k, h) = (3, 1)$ and axis of symmetry is $y = 1$. If $y = 0$ then $x = 2(-1)^2 + 3 = 5$ and x-intercept is $(5, 0)$. If $x = 0$, we obtain $2(y - 1)^2 + 3 = 0$ which is inconsistent since the left-hand side is always positive. No y-intercept.

Since $\dfrac{1}{4p} = a = 2$, $p = 1/8$, focus is $(k + p, h) = (25/8, 1)$, and directrix is $x = k - p = 23/8$.

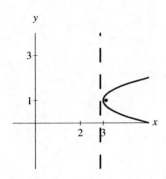

71. Since $x = -\dfrac{1}{2}(y+2)^2 + 1$ is of the form $a(y-h)^2 + k$, vertex is $(k,h) = (1,-2)$ and axis of symmetry is $y = -2$. If $y = 0$, then $x = -\dfrac{1}{2}(2)^2 + 1 = -1$ and x-intercept is $(-1,0)$. If $x = 0$, then

$$
\begin{aligned}
\frac{1}{2}(y+2)^2 &= 1 \\
(y+2)^2 &= 2 \\
y &= -2 \pm \sqrt{2}.
\end{aligned}
$$

The y-intercepts are $(0, -2 \pm \sqrt{2})$. Since $\dfrac{1}{4p} = a = -\dfrac{1}{2}$, we find $p = -\dfrac{1}{2}$, focus is $(k+p,h) = (1/2,-2)$, and directrix is $x = k - p = 3/2$.

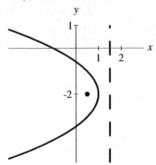

73. Since focus is 1 unit above the vertex $(1,4)$, we obtain $p = 1$. Then $a = \dfrac{1}{4p} = \dfrac{1}{4}$ and parabola is given by $y = \dfrac{1}{4}(x-1)^2 + 4$.

75. Since vertex $(0,0)$ is 2 units to the right of the directrix, we find $p = 2$ and parabola opens to the right. Then $a = \dfrac{1}{4p} = \dfrac{1}{8}$ and the parabola is given by $x = \dfrac{1}{8}y^2$.

77. Since the parabola opens up, $p = 55(12)$ inches, and the vertex is $(0,0)$. Note, $\dfrac{1}{4p} = \dfrac{1}{2640}$. Thus, the parabola is given by $y = \dfrac{1}{2640}x^2$. Thickness at the outside edge is $23 + \dfrac{1}{2640}(100)^2 \approx 26.8$ in.

79. Let $y = ax^2$ be an equation of the parabola. Since $(3264, 675)$ lies on the parabola, we get

$$
\begin{aligned}
675 &= a(3264)^2 \\
a &= \frac{675}{3264^2}.
\end{aligned}
$$

Thus, an equation of the parabola is

$$
y = \frac{675}{3264^2}x^2
$$

or

$$
y \approx 6.3 \times 10^{-5}x^2.
$$

Note, the coordinates of the focus of the parabola is

$$
\left(0, \frac{1}{4a}\right)
$$

which is about

$$
(0, 3946).
$$

For Thought

1. False, y-intercepts are $(\pm 3, 0)$.

2. True, since it can be written as $\dfrac{x^2}{1/2} + y^2 = 1$.

3. True, length of the major axis is $2a = 2(5) = 10$.

4. True, if $y = 0$ then $x^2 = \dfrac{1}{0.5} = 2$ and $x = \pm\sqrt{2}$.

5. True, if $x = 0$ then $y^2 = 3$ and $y = \pm\sqrt{3}$.

6. False, the center is not a point on the circle.

7. True   **8.**   False, $(3,-1)$ satisfies equation.

9. False. No point satisfies the equation since the left-hand side is always positive.

10. False, since the circle can be written as $(x-2)^2 + (y+1/2)^2 = 53/4$ and so the radius is $\sqrt{53}/2$.

5.2 Exercises

1. Foci $(\pm\sqrt{5}, 0)$, vertices $(\pm 3, 0)$, center $(0,0)$

3. Foci $(2, 1 \pm \sqrt{5})$, vertices $(2,4)$ and $(2,-2)$, center $(2,1)$

5. Since $c = 2$ and $b = 3$, we get
$$a^2 = b^2 + c^2 = 9 + 4 = 13 \text{ and } a = \sqrt{13}.$$
Ellipse is given by $\dfrac{x^2}{13} + \dfrac{y^2}{9} = 1.$

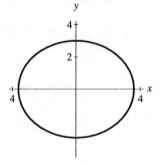

7. Since $c = 4$ and $a = 5$, we find
$$b^2 = a^2 - c^2 = 25 - 16 = 9 \text{ and } b = 3.$$
Ellipse is given by $\dfrac{x^2}{25} + \dfrac{y^2}{9} = 1.$

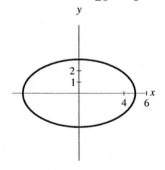

9. Since $c = 2$ and $b = 2$, we obtain
$$a^2 = b^2 + c^2 = 4 + 4 = 8 \text{ and } a = \sqrt{8}.$$
Ellipse is given by $\dfrac{x^2}{4} + \dfrac{y^2}{8} = 1.$

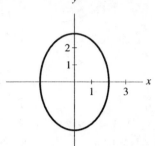

11. Since $c = 4$ and $a = 7$, we obtain
$$b^2 = a^2 - c^2 = 49 - 16 = 33 \text{ and } b = \sqrt{33}.$$
Ellipse is given by $\dfrac{x^2}{33} + \dfrac{y^2}{49} = 1.$

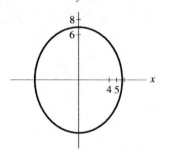

13. Since $c = \sqrt{a^2 - b^2} = \sqrt{16 - 4} = 2\sqrt{3}$, the foci are $(\pm 2\sqrt{3}, 0)$

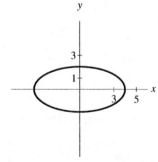

15. Since $c = \sqrt{a^2 - b^2} = \sqrt{36 - 9} = \sqrt{27}$, the foci are $(0, \pm 3\sqrt{3})$

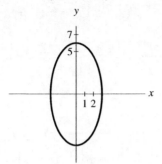

17. Since $c = \sqrt{a^2 - b^2} = \sqrt{25 - 1} = \sqrt{24}$, the foci are $(\pm 2\sqrt{6}, 0)$

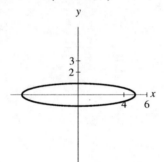

19. Since $c = \sqrt{a^2 - b^2} = \sqrt{25 - 9} = \sqrt{16}$, the foci are $(0, +4)$

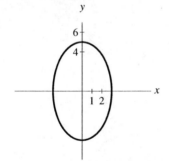

21. From $x^2 + \dfrac{y^2}{9} = 1$, one finds $c = \sqrt{a^2 - b^2}$
$= \sqrt{9 - 1} = \sqrt{8}$ and foci are $(0, \pm 2\sqrt{2})$.

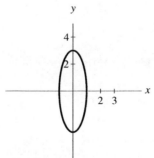

23. From $\dfrac{x^2}{9} + \dfrac{y^2}{4} = 1$, one finds $c = \sqrt{a^2 - b^2}$
$= \sqrt{9 - 4} = \sqrt{5}$ and foci are $(\pm\sqrt{5}, 0)$.

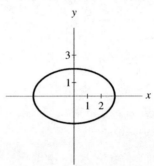

25. Since $c = \sqrt{a^2 - b^2} = \sqrt{16 - 9} = \sqrt{7}$, the foci are $(1 \pm c, -3) = (1 \pm \sqrt{7}, -3)$.

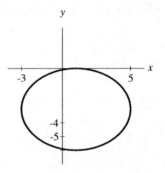

27. Since $c = \sqrt{a^2 - b^2} = \sqrt{25 - 9} = \sqrt{16} = 4$, the foci are $(3, -2 \pm c)$, or $(3, 2)$ and $(3, -6)$.

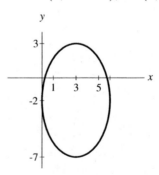

29. Since $\dfrac{(x+4)^2}{36} + (y+3)^2 = 1$, we get

$$c = \sqrt{a^2 - b^2} = \sqrt{36 - 1} = \sqrt{35}$$

and the foci are

$$(-4 \pm \sqrt{35}, -3).$$

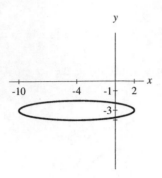

35. Since $\dfrac{x^2}{16} + \dfrac{y^2}{4} = 1$, we get $c = \sqrt{a^2 - b^2}$

$= \sqrt{16 - 4} = \sqrt{12}$, and the foci are $(\pm 2\sqrt{3}, 0)$.

37. Since $\dfrac{(x+1)^2}{4} + \dfrac{(y+2)^2}{16} = 1$, we get

$c = \sqrt{a^2 - b^2} = \sqrt{16 - 4} = \sqrt{12}$, and the foci are $(-1, -2 \pm 2\sqrt{3})$.

31. If one applies the method of completing the square, one obtains

$$
\begin{aligned}
9(x^2 - 2x + 1) + 4(y^2 + 4y + 4) &= 11 + 9 + 16 \\
9(x - 1)^2 + 4(y + 2)^2 &= 36 \\
\frac{(x - 1)^2}{4} + \frac{(y + 2)^2}{9} &= 1.
\end{aligned}
$$

From $a^2 = b^2 + c^2$ with $a = 3$ and $b = 2$, one finds $c = \sqrt{5}$. The foci are $(1, -2 \pm \sqrt{5})$ and a sketch of the ellipse is given.

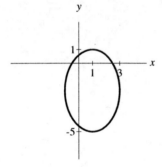

39. Since $r = \sqrt{(4 - 0)^2 + (5 - 0)^2} = \sqrt{41}$, the circle is given by $x^2 + y^2 = 41$.

41. Since $r = \sqrt{(4 - 2)^2 + (1 + 3)^2} = \sqrt{20}$, the circle is given by $(x - 2)^2 + (y + 3)^2 = 20$.

43. Since center is $\left(\dfrac{3 - 1}{2}, \dfrac{4 + 2}{2}\right) = (1, 3)$ and

$r = \sqrt{(3 - 1)^2 + (4 - 3)^2} = \sqrt{5}$, the circle is given by $(x - 1)^2 + (y - 3)^2 = 5$.

45. center $(0, 0)$, radius 10

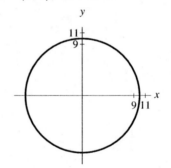

33. Applying the method of completing the square, we find

$$
\begin{aligned}
9(x^2 - 6x + 9) + 4(y^2 + 4y + 4) &= -61 + 81 + 16 \\
9(x - 3)^2 + 4(y + 2)^2 &= 36 \\
\frac{(x - 3)^2}{4} + \frac{(y + 2)^2}{9} &= 1.
\end{aligned}
$$

From $a^2 = b^2 + c^2$ with $a = 3$ and $b = 2$, we find $c = \sqrt{5}$. The foci are $(3, -2 \pm \sqrt{5})$ and a sketch of the ellipse is given.

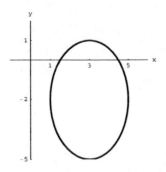

47. center $(1, 2)$, radius 2

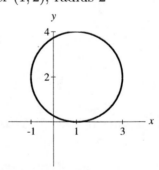

49. Completing the square, we obtain

$$
\begin{aligned}
x^2 + (y^2 + 2y + 1) &= 8 + 1 \\
x^2 + (y + 1)^2 &= 9.
\end{aligned}
$$

The center is $(0, -1)$ with radius 3.

51. Completing the square, we find

$$x^2 + 8x + 16 + y^2 - 10y + 25 = 16 + 25$$
$$(x + 4)^2 + (y - 5)^2 = 41.$$

The center is $(-4, 5)$ with radius $\sqrt{41}$.

53. Completing the square, we find

$$(x^2 + 4x + 4) + y^2 = 5 + 4$$
$$(x + 2)^2 + y^2 = 9.$$

The center is $(-2, 0)$ with radius 3.

55. Completing the square, we have

$$x^2 - x + \frac{1}{4} + y^2 + y + \frac{1}{4} = \frac{1}{2} + \frac{1}{4} + \frac{1}{4}$$
$$(x - 1/2)^2 + (y + 1/2)^2 = 1.$$

The center is $(0.5, -0.5)$ with radius 1.

57. Completing the square, we get

$$y^2 - y + x^2 = 0$$
$$y^2 - y + \frac{1}{4} + x^2 = \frac{1}{4}$$
$$(y - 1/2)^2 + x^2 = \frac{1}{4},$$

which is a circle.

59. Divide equation by 4.

$$x^2 + 3y^2 = 1$$
$$x^2 + \frac{y^2}{1/3} = 1$$

This is an ellipse.

61. Solve for y and complete the square.

$$y = -2x^2 - 4x + 4$$
$$y = -2(x^2 + 2x) + 4$$
$$y = -2(x^2 + 2x + 1) + 4 + 2$$
$$y = -2(x + 1)^2 + 6$$

We find a parabola.

63. Note, $(y - 2)^2 = (2 - y)^2$. Then we solve for x.

$$2 - x = (y - 2)^2$$
$$x = -(y - 2)^2 + 2$$

This is a parabola.

65. Simplify and note $(x - 4)^2 = (4 - x)^2$.

$$2(x - 4)^2 = 4 - y^2$$
$$2(x - 4)^2 + y^2 = 4$$
$$\frac{(x - 4)^2}{2} + \frac{y^2}{4} = 1$$

We find an ellipse.

67. Divide given equation by 9 to get the circle given by $x^2 + y^2 = \frac{1}{9}$.

69. From the foci, we obtain $c = 2$ and $a^2 = b^2 + c^2 = b^2 + 4$. Equation of the ellipse is of the form $\frac{x^2}{b^2 + 4} + \frac{y^2}{b^2} = 1$. Substitute $x = 2$ and $y = 3$.

$$\frac{4}{b^2 + 4} + \frac{9}{b^2} = 1$$
$$4b^2 + 9(b^2 + 4) = b^2(b^2 + 4)$$
$$0 = b^4 - 9b^2 - 36$$
$$0 = (b^2 - 12)(b^2 + 3)$$

So $b^2 = 12$ and $a^2 = 12 + 4 = 16$.

The ellipse is given by $\frac{x^2}{16} + \frac{y^2}{12} = 1$.

71. If c is the distance between the center and focus $(0, 0)$ then the other focus is $(2c, 0)$. Since the distance between the x-intercepts is $6 + 2c$, which is also the length of the major axis, then $6 + 2c = 2a$. Since $2a$ is the sum of the distances of $(0, 5)$ from the foci,

$$5 + \sqrt{25 + 4c^2} = 2a$$
$$5 + \sqrt{25 + 4c^2} = 6 + 2c$$
$$\sqrt{25 + 4c^2} = 1 + 2c$$
$$25 + 4c^2 = 1 + 4c + 4c^2$$
$$24 = 4c$$
$$6 = c.$$

The other focus is $(2c, 0) = (12, 0)$.

73. Since the sun is a focus of the elliptical orbit, the length of the major axis is $2a = 521$ (the sum of the shortest distance, $P = 1$ AU, and longest distance, $A = 520$ AU, between the orbit and the sun, respectively). In addition, $c = 259.5$ AU (which is the distance from the center to a focus). The eccentricity is

$$e = \frac{c}{a} = \frac{259.5}{260.5} \approx 0.996.$$

Orbit's equation is

$$\frac{x^2}{260.5^2} + \frac{y^2}{520} = 1.$$

75. If $2a$ is the sum of the distances from Haley's comet to the two foci and c is the distance from the sun to the center of the ellipse then, $c = a - 8 \times 10^7$.

Since $0.97 = c/a$, we get

$$0.97 = \frac{a - 8 \times 10^7}{a}$$

and the solution of this equation is $a \approx 2.667 \times 10^9$. So $c = a(0.97) \approx 2.587 \times 10^9$. The maximum distance from the sun is $c + a \approx 5.25 \times 10^9$ km.

For Thought

1. False, it is a parabola.

2. False, there is no y-intercept.

3. True   **4.**   True

5. False, $y = \frac{b}{a}x$ is an asymptote.

6. True   **7.**   True, $c = \sqrt{16 + 9} = 5$.

8. True, $c = \sqrt{3 + 5} = \sqrt{8}$.

9. False, $y = \frac{2}{3}x$ is an asymptote.

10. False, it is a circle centered at $(0,0)$.

5.3 Exercises

1. Vertices $(\pm 1, 0)$, foci $(\pm\sqrt{2}, 0)$, asymptotes $y = \pm x$

3. Vertices $(1, \pm 3)$, foci $(1, \pm\sqrt{10})$

Since the slope of the asymptotes are ± 3, the equations of the asymptotes are of the form $y = \pm 3x + b$. If we substitute $(1, 0)$ into $y = \pm 3x + b$, then $0 = \pm 3(1) + b$ or $b = \mp 3$. Thus, the asymptotes are $y = 3x - 3$ and $y = -3x + 3$.

5. Note, $c = \sqrt{a^2 + b^2} = \sqrt{2^2 + 3^2} = \sqrt{13}$.

Foci $(\pm\sqrt{13}, 0)$, asymptotes $y = \pm\frac{b}{a}x = \pm\frac{3}{2}x$

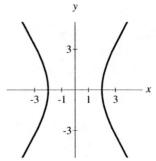

7. Note, $c = \sqrt{a^2 + b^2} = \sqrt{2^2 + 5^2} = \sqrt{29}$.

Foci $(0, \pm\sqrt{29})$, asymptotes $y = \pm\frac{a}{b}x = \pm\frac{2}{5}x$

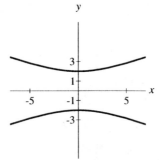

9. Note, $c = \sqrt{a^2 + b^2} = \sqrt{2^2 + 1^2} = \sqrt{5}$.

Foci $(\pm\sqrt{5}, 0)$, asymptotes $y = \pm\frac{b}{a}x = \pm\frac{1}{2}x$

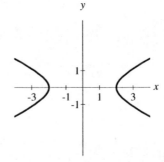

11. Note, $c = \sqrt{a^2 + b^2} = \sqrt{1^2 + 3^2} = \sqrt{10}$.

Foci $(\pm\sqrt{10}, 0)$, asymptotes $y = \pm\dfrac{b}{a}x = \pm 3x$

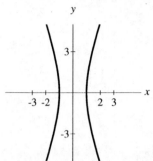

13. Dividing by 144, we get $\dfrac{x^2}{9} - \dfrac{y^2}{16} = 1$.

Note, $c = \sqrt{a^2 + b^2} = \sqrt{3^2 + 4^2} = 5$.

Foci $(\pm 5, 0)$, asymptotes $y = \pm\dfrac{b}{a}x = \pm\dfrac{4}{3}x$

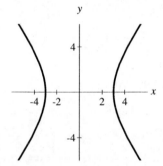

15. Note, $c = \sqrt{a^2 + b^2} = \sqrt{1^2 + 1^2} = \sqrt{2}$.

Foci $(\pm\sqrt{2}, 0)$, asymptotes $y = \pm\dfrac{b}{a}x = \pm x$

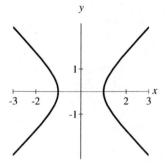

17. Note, $c = \sqrt{a^2 + b^2} = \sqrt{2^2 + 3^2} = \sqrt{13}$.
Since the center is $(-1, 2)$, we find that the foci are $(-1 \pm \sqrt{13}, 2)$. Solving for y in $y - 2 = \pm\dfrac{3}{2}(x + 1)$, we obtain that the asymptotes are $y = \dfrac{3}{2}x + \dfrac{7}{2}$ and $y = -\dfrac{3}{2}x + \dfrac{1}{2}$.

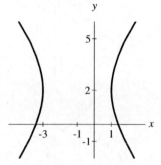

19. Note, $c = \sqrt{a^2 + b^2} = \sqrt{2^2 + 1^2} = \sqrt{5}$.
Since the center is $(-2, 1)$, the foci are $(-2, 1 \pm \sqrt{5})$. Solving for y in $y - 1 = \pm 2(x + 2)$, we obtain that the asymptotes are $y = 2x + 5$ and $y = -2x - 3$.

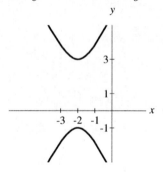

21. Note, $c = \sqrt{a^2 + b^2} = \sqrt{4^2 + 3^2} = 5$.
Since the center is $(-2, 3)$, the foci are $(3, 3)$ and $(-7, 3)$. Solving for y in $y - 3 = \pm\dfrac{3}{4}(x + 2)$, we find that the asymptotes are $y = \dfrac{3}{4}x + \dfrac{9}{2}$ and $y = -\dfrac{3}{4}x + \dfrac{3}{2}$.

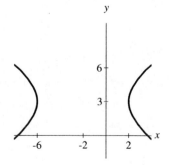

23. Note, $c = \sqrt{a^2 + b^2} = \sqrt{1^2 + 1^2} = \sqrt{2}$.
Since the center is $(3, 3)$, the foci
are $(3, 3 \pm \sqrt{2})$. Solving for y in
$y - 3 = \pm(x - 3)$, we obtain that the asymptotes
are $y = x$ and $y = -x + 6$.

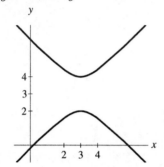

25. Since the x-intercepts are $(\pm 6, 0)$, the
hyperbola is given by $\dfrac{x^2}{6^2} - \dfrac{y^2}{b^2} = 1$.
From the asymptotes one gets $\dfrac{1}{2} = \dfrac{b}{6}$
and $b = 3$. An equation is $\dfrac{x^2}{36} - \dfrac{y^2}{9} = 1$.

27. Since the x-intercepts are $(\pm 3, 0)$, the
hyperbola is given by $\dfrac{x^2}{3^2} - \dfrac{y^2}{b^2} = 1$.
From the foci, $c = 5$ and $b^2 = c^2 - a^2$
$= 5^2 - 3^2 = 16$. An equation is $\dfrac{x^2}{9} - \dfrac{y^2}{16} = 1$.

29. By using the vertices of the fundamental
rectangle and since it opens sideways,
one gets $a = 3$, $b = 5$, and the center is
at the origin. An equation is $\dfrac{x^2}{9} - \dfrac{y^2}{25} = 1$.

31. $\dfrac{x^2}{9} - \dfrac{y^2}{16} = 1$

33. $\dfrac{y^2}{9} - \dfrac{(x - 1)^2}{9} = 1$

35. Completing the square,

$$\begin{aligned}
y^2 - (x^2 - 2x) &= 2 \\
y^2 - (x^2 - 2x + 1) &= 2 - 1 \\
y^2 - (x - 1)^2 &= 1
\end{aligned}$$

we obtain a hyperbola.

37. $y = x^2 + 2x$ is a parabola.

39. Simplifying,

$$\begin{aligned}
25x^2 + 25y^2 &= 2500 \\
x^2 + y^2 &= 100
\end{aligned}$$

we obtain a circle.

41. Simplifying,

$$\begin{aligned}
25x &= -100y^2 + 2500 \\
x &= -4y^2 + 100
\end{aligned}$$

we find a parabola.

43. Completing the square,

$$\begin{aligned}
2(x^2 - 2x) + 2(y^2 - 4y) &= -9 \\
2(x^2 - 2x + 1) + 2(y^2 - 4y + 4) &= -9 + 2 + 8 \\
2(x - 1)^2 + 2(y - 2)^2 &= 1 \\
(x - 1)^2 + (y - 2)^2 &= \dfrac{1}{2}
\end{aligned}$$

we find a circle.

45. Completing the square,

$$\begin{aligned}
2(x^2 + 2x) + y^2 + 6y &= -7 \\
2(x^2 + 2x + 1) + y^2 + 6y + 9 &= -7 + 2 + 9 \\
2(x + 1)^2 + (y + 3)^2 &= 4 \\
\dfrac{(x + 1)^2}{2} + \dfrac{(y + 3)^2}{4} &= 1
\end{aligned}$$

we get an ellipse.

47. Completing the square, we find

$$\begin{aligned}
25(x^2 - 6x + 9) - 4(y^2 + 2y + 1) &= -121 + 225 - 4 \\
25(x - 3)^2 - 4(y + 1)^2 &= 100 \\
\dfrac{(x - 3)^2}{4} - \dfrac{(y + 1)^2}{25} &= 1.
\end{aligned}$$

We have a hyperbola.

49. Ellipse, since $AC = (4)(3) > 0$ and $A \neq C$.

51. Parabola, since $AC = (2)(0) = 0$.

53. Hyperbola, since $AC = (9)(-5) < 0$.

55. From the center $(0, 0)$ and vertex $(0, 8)$ one
gets $a = 8$. From the foci $(0, \pm 10)$, $c = 10$.
So $b^2 = c^2 - a^2 = 10^2 - 8^2 = 36$. The
hyperbola is given by $\dfrac{y^2}{64} - \dfrac{x^2}{36} = 1$.

57. Multiply $16y^2 - x^2 = 16$ by 9 and add to $9x^2 - 4y^2 = 36$.

$$
\begin{array}{rcl}
-9x^2 + 144y^2 &=& 144 \\
9x^2 - 4y^2 &=& 36 \\
\hline
140y^2 &=& 180 \\
y^2 &=& \dfrac{9}{7} \\
y &=& \dfrac{3\sqrt{7}}{7}
\end{array}
$$

Using $y^2 = \dfrac{9}{7}$ in $x^2 = 16(y^2 - 1)$, we get

$$x^2 = 16\left(\dfrac{2}{7}\right) \text{ or } x = \dfrac{4\sqrt{14}}{7}.$$

The exact location is $\left(\dfrac{4\sqrt{14}}{7}, \dfrac{3\sqrt{7}}{7}\right)$.

59. Since $c^2 = a^2 + b^2 = 1^2 + 1^2 = 2$, the foci of $x^2 - y^2 = 1$ are $A(\sqrt{2}, 0)$ and $B(-\sqrt{2}, 0)$. Note, $y^2 = x^2 - 1$. Suppose (x, y) is a point on the hyperbola whose distance from B is twice the distance between (x, y) and A. Then

$$
\begin{array}{rcl}
2\sqrt{(x - \sqrt{2})^2 + y^2} &=& \sqrt{(x + \sqrt{2})^2 + y^2} \\
4((x - \sqrt{2})^2 + y^2) &=& (x + \sqrt{2})^2 + y^2
\end{array}
$$

$$
\begin{array}{rcl}
4(x - \sqrt{2})^2 - (x + \sqrt{2})^2 + 3y^2 &=& 0 \\
3x^2 - 10\sqrt{2}x + 6 + 3y^2 &=& 0 \\
3x^2 - 10\sqrt{2}x + 6 + 3(x^2 - 1) &=& 0 \\
6x^2 - 10\sqrt{2}x + 3 &=& 0.
\end{array}
$$

Solving for x, one finds $x = \dfrac{3\sqrt{2}}{2}$ and $x = \dfrac{\sqrt{2}}{6}$; the second value must be excluded since it is out of the domain. Substituting $x = \dfrac{3\sqrt{2}}{2}$ into $y^2 = x^2 - 1$, one obtains $y = \pm\dfrac{\sqrt{14}}{2}$.

By the symmetry of the hyperbola, there are four points that are twice as far from one focus as they are from the other focus. Namely, these points are

$$\left(\pm\dfrac{3\sqrt{2}}{2}, \pm\dfrac{\sqrt{14}}{2}\right).$$

61. Note, the asymptotes are $y = \pm x$. The difference is

$$50 - \sqrt{50^2 - 1} \approx 0.01$$

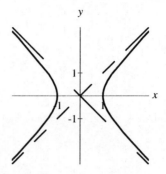

For Thought

1. True   **2.**   False, the distance is $|r|$.

3. False

4. False, $x = r\cos\theta$, $y = r\sin\theta$, and $x^2 + y^2 = r^2$.

5. True, since $x = -4\cos 225° = 2\sqrt{2}$ and $y = -4\sin 225° = 2\sqrt{2}$.

6. True, $\theta = \pi/4$ is a straight line through the origin which makes an angle of $\pi/4$ with the positive x-axis.

7. True, since each circle is centered at the origin with radius 5.

8. False, since upon substitution one gets $\cos 2\pi/3 = -1/2$ while $r^2 = 1/2$.

9. False, for $r = 1/\sin\theta$ is undefined when $\theta = k\pi$ for any integer k.

10. False, since $r = \theta$ is a reflection of $r = -\theta$ about the origin.

5.4 Exercises

1. $(3, \pi/2)$

3. Since $r = \sqrt{3^2 + 3^2} = 3\sqrt{2}$ and $\theta = \tan^{-1}(3/3) = \pi/4$, $(r, \theta) = (3\sqrt{2}, \pi/4)$.

5. $(2, 0°)$

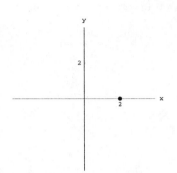

7. $(0, 35°)$

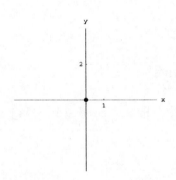

9. $(3, \pi/6)$

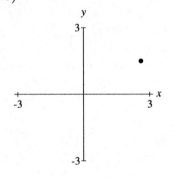

11. $(-2, 2\pi/3)$

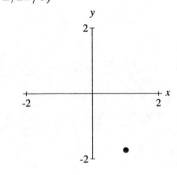

13. $(2, -\pi/4)$

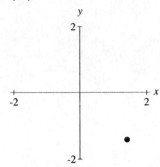

15. $(3, -225°)$

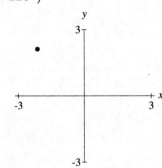

17. $(-2, 45°)$

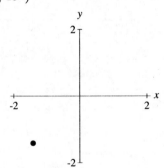

19. $(4, 390°)$

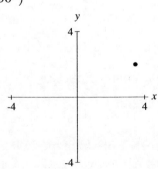

21. $(x, y) = (4 \cdot \cos(0°), 4 \cdot \sin(0°)) = (4, 0)$

23. $(x, y) = (0 \cdot \cos(\pi/4), 0 \cdot \sin(\pi/4)) = (0, 0)$

25. $(x, y) = (1 \cdot \cos(\pi/6), 1 \cdot \sin(\pi/6)) =$

$$\left(\frac{\sqrt{3}}{2}, \frac{1}{2}\right)$$

27. $(x, y) = (-3\cos(3\pi/2), -3\sin(3\pi/2)) = (0, 3)$

29. $(x, y) = \left(\sqrt{2}\cos 135°, \sqrt{2}\sin 135°\right) = (-1, 1)$

31. $(x, y) = \left(-\sqrt{6}\cos(-60°), -\sqrt{6}\sin(-60°)\right) =$
$$\left(-\frac{\sqrt{6}}{2}, \frac{3\sqrt{2}}{2}\right)$$

33. Since $r = \sqrt{(\sqrt{3})^2 + 3^2} = 2\sqrt{3}$ and
$$\cos\theta = \frac{x}{r} = \frac{\sqrt{3}}{2\sqrt{3}} = \frac{1}{2}, \text{ one can choose}$$
$\theta = 60°$. So $(r, \theta) = (2\sqrt{3}, 60°)$.

35. Since $r = \sqrt{(-2)^2 + 2^2} = 2\sqrt{2}$ and
$$\cos\theta = \frac{x}{r} = \frac{-2}{2\sqrt{2}} = -\frac{1}{\sqrt{2}}, \text{ one can choose}$$
$\theta = 135°$. So $(r, \theta) = (2\sqrt{2}, 135°)$.

37. $(r, \theta) = (2, 90°)$

39. Note $r = \sqrt{(-3)^2 + (-3)^2} = 3\sqrt{2}$.
Since $\tan\theta = \frac{y}{x} = \frac{-3}{-3} = 1$ and $(-3, -3)$
is in quadrant III, one can choose $\theta = 225°$.
Thus, $(r, \theta) = (3\sqrt{2}, 225°)$.

41. Since $r = \sqrt{1^2 + 4^2} = \sqrt{17}$ and
$$\theta = \cos^{-1}\left(\frac{x}{r}\right) = \cos^{-1}\left(\frac{1}{\sqrt{17}}\right) \approx 75.96°$$
one gets $(r, \theta) = (\sqrt{17}, 75.96°)$.

43. Since $r = \sqrt{(\sqrt{2})^2 + (-2)^2} = \sqrt{6}$ and
$$\theta = \tan^{-1}\left(\frac{y}{x}\right) = \tan^{-1}\left(\frac{-2}{\sqrt{2}}\right) \approx -54.7°,$$
one obtains $(r, \theta) = (\sqrt{6}, -54.7°)$.

45. $(x, y) = (4\cos 26°, 4\sin 26°) \approx (3.60, 1.75)$

47. $(x, y) = (2\cos(\pi/7), 2\sin(\pi/7)) \approx (1.80, 0.87)$

49. $(x, y) = (-2\cos(1.1), -2\sin(1.1)) \approx (-0.91, -1.78)$

51. Since $r = \sqrt{4^2 + 5^2} \approx 6.4$ and
$\tan\theta = \frac{y}{x} = \frac{5}{4}$, we get $\theta = \tan^{-1}\left(\frac{5}{4}\right) \approx$
$51.3°$. Then $(r, \theta) \approx (6.4, 51.3°)$.

53. Note, $r = \sqrt{(-2)^2 + (-7)^2} \approx 7.3$ and
$\tan^{-1}\left(\frac{-7}{-2}\right) \approx 74.1°$. Since $(-2, -7)$ is
a point in the 3rd quadrant, we choose
$\theta = 74.1° + 180° = 254.1°$.
Then $(r, \theta) \approx (7.3, 254.1°)$.

55. $r = 2\sin\theta$ is a circle centered at $(x, y) = (0, 1)$.
It goes through the following points in polar
coordinates: $(0, 0)$, $(1, \pi/6)$, $(2, \pi/2)$,
$(1, 5\pi/6)$, $(0, \pi)$.

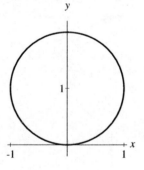

57. $r = 3\cos 2\theta$ is a four-leaf rose that goes
through the following points in polar
coordinates $(3, 0)$, $(3/2, \pi/6)$,
$(0, \pi/4)$, $(-3, \pi/2)$, $(0, 3\pi/4)$, $(3, \pi)$,
$(-3, 3\pi/2)$, $(0, 7\pi/4)$.

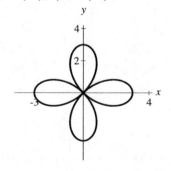

59. $r = 2\theta$ is spiral-shaped and goes through the following points in polar coordinates $(-\pi, -\pi/2)$, $(0, 0)$, $(\pi, \pi/2)$, $(2\pi, \pi)$

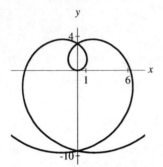

61. $r = 1 + \cos\theta$ goes through the following points in polar coordinates $(2, 0)$, $(1.5, \pi/3)$, $(1, \pi/2)$, $(0.5, 2\pi/3)$, $(0, \pi)$.

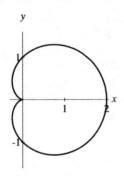

63. $r^2 = 9\cos 2\theta$ goes through the following points in polar coordinates $(0, -\pi/4)$, $(\pm\sqrt{3}/2, -\pi/6)$, $(\pm 3, 0)$, $(\pm\sqrt{3}/2, \pi/6)$, $(0, \pi/4)$.

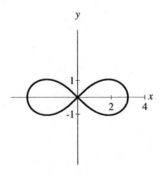

65. $r = 4\cos 2\theta$ is a four-leaf rose that goes through the following points in polar coordinates $(4, 0)$, $(2, \pi/6)$, $(0, \pi/4)$, $(-4, \pi/2)$, $(2, 5\pi/6)$, $(4, \pi)$, $(0, 5\pi/4)$, $(-4, 3\pi/2)$, $(0, 7\pi/4)$.

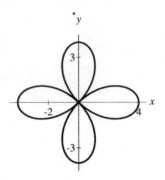

67. $r = 2\sin 3\theta$ is a three-leaf rose that goes through the following points in polar coordinates $(0, 0)$, $(2, \pi/6)$, $(-2, \pi/2)$, $(2, 5\pi/6)$.

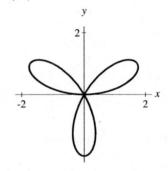

69. $r = 1 + 2\cos\theta$ goes through the following points in polar coordinates $(3, 0)$, $(2, \pi/3)$, $(1, \pi/2)$, $(0, 2\pi/3)$, $(1 - \sqrt{3}, 5\pi/6)$, $(-1, \pi)$.

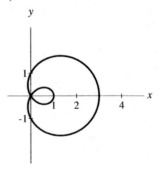

71. $r = 3.5$ is a circle centered at the origin with radius 3.5.

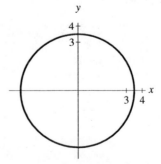

73. $\theta = 30°$ is a line through the origin that makes a 30° angle with the positive x-axis.

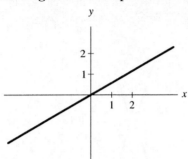

75. Multiply equation by r.

$$
\begin{aligned}
r^2 &= 4r\cos\theta \\
x^2 + y^2 &= 4x \\
x^2 - 4x + y^2 &= 0
\end{aligned}
$$

77. Multiply equation by $\sin\theta$.

$$
\begin{aligned}
r\sin\theta &= 3 \\
y &= 3
\end{aligned}
$$

79. Multiply equation by $\cos\theta$.

$$
\begin{aligned}
r\cos\theta &= 3 \\
x &= 3
\end{aligned}
$$

81. Since $r = 5$, $\sqrt{x^2 + y^2} = 5$ and by squaring one gets $x^2 + y^2 = 25$.

83. Note $\tan\theta = \dfrac{\sin\theta}{\cos\theta} = \dfrac{y/r}{x/r} = \dfrac{y}{x}$.
Since $\tan\pi/4 = 1$, $\dfrac{y}{x} = 1$ or $y = x$.

85. Multiply equation by $1 - \sin\theta$.

$$
\begin{aligned}
r(1 - \sin\theta) &= 2 \\
r - r\sin\theta &= 2 \\
r - y &= 2 \\
\pm\sqrt{x^2 + y^2} &= y + 2 \\
x^2 + y^2 &= y^2 + 4y + 4 \\
x^2 - 4y &= 4
\end{aligned}
$$

87. $r\cos\theta = 4$

89. Note $\tan\theta = y/x$.

$$
\begin{aligned}
y &= -x \\
\frac{y}{x} &= -1 \\
\tan\theta &= -1 \\
\theta &= -\pi/4
\end{aligned}
$$

91. Note $x = r\cos\theta$ and $y = r\sin\theta$.

$$
\begin{aligned}
(r\cos\theta)^2 &= 4r\sin\theta \\
r^2\cos^2\theta &= 4r\sin\theta \\
r &= \frac{4\sin\theta}{\cos^2\theta} \\
r &= 4\tan\theta\sec\theta
\end{aligned}
$$

93. $r = 2$

95. Note $x = r\cos\theta$ and $y = r\sin\theta$.

$$
\begin{aligned}
y &= 2x - 1 \\
r\sin\theta &= 2r\cos\theta - 1 \\
r(\sin\theta - 2\cos\theta) &= -1 \\
r(2\cos\theta - \sin\theta) &= 1 \\
r &= \frac{1}{2\cos\theta - \sin\theta}
\end{aligned}
$$

97. Note that $y = r\sin\theta$ and $x^2 + y^2 = r^2$.

$$
\begin{aligned}
x^2 + (y^2 - 2y + 1) &= 1 \\
x^2 + y^2 - 2y &= 0 \\
r^2 - 2r\sin\theta &= 0 \\
r^2 &= 2r\sin\theta \\
r &= 2\sin\theta
\end{aligned}
$$

For Thought

1. True

2. False, rather the eccentricity satisfies $0 < e < 1$.

3. True   **4.** True   **5.** True

6. False. Since $r = \dfrac{\frac{1}{3}\cdot 4}{1 - \frac{1}{3}\cos\theta}$, the directrix is $x = -4$.   **7.** True

8. True, for the discriminant $B^2 - 4AC = 0^2 - 4(2)(-3)$ is positive.

9. True, for the discriminant $B^2 - 4AC = 0^2 - 4(2)(0)$ is zero.

10. True, for the discriminant $B^2 - 4AC = 0^2 - 4(2)(5)$ is negative.

5.5 Exercises

1. Since $r = \dfrac{2 \cdot 3}{1 - 2\cos\theta}$, eccentricity is $e = 2$, conic is a hyperbola, and distance is $p = 3$.

3. Since $r = \dfrac{1 \cdot \frac{3}{4}}{1 - \sin\theta}$, eccentricity is $e = 1$, conic is a parabola, and distance is $p = \dfrac{3}{4}$.

5. Since $r = \dfrac{\frac{4}{3} \cdot \frac{3}{4}}{1 + \frac{4}{3}\sin\theta}$, eccentricity is $e = \dfrac{4}{3}$, conic is a hyperbola, and distance is $p = \dfrac{3}{4}$.

7. A parabola with directrix $y = -2$

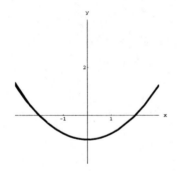

9. An ellipse with directrix $x = \dfrac{5}{2}$ since

$$r = \dfrac{\frac{2}{3} \cdot \frac{5}{2}}{1 + \frac{2}{3}\cos\theta}$$

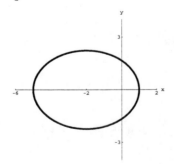

11. A hyperbola with directrix $x = -\dfrac{1}{6}$ since

$$r = \dfrac{3 \cdot \frac{1}{6}}{1 - 3\cos\theta}$$

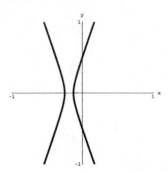

13. An ellipse with directrix $y = 6$ since

$$r = \dfrac{\frac{1}{2} \cdot 6}{1 + \frac{1}{2}\sin\theta}$$

15. $r = \dfrac{2}{1 + \sin\theta}$

17. Since $r = \dfrac{2 \cdot 5}{1 + 2\cos\theta}$, we find $r = \dfrac{10}{1 + 2\cos\theta}$.

19. Note, $r = \dfrac{\frac{1}{2} \cdot 4}{1 + \frac{1}{2}\sin\theta}$. So $r = \dfrac{4}{2 + \sin\theta}$.

21. Note, $r = \dfrac{\frac{3}{4} \cdot 8}{1 - \frac{3}{4}\cos\theta}$. Then $r = \dfrac{24}{4 - 3\cos\theta}$.

23. Multiplying by $1 + \sin\theta$, we obtain

$$
\begin{aligned}
r + r\sin\theta &= 3 \\
r &= 3 - y \\
x^2 + y^2 &= 9 - 6y + y^2.
\end{aligned}
$$

Simplifying, we get $x^2 + 6y - 9 = 0$ is a parabola.

25. Multiplying by $4 - \cos\theta$, we get

$$
\begin{aligned}
4r - r\cos\theta &= 3 \\
4r &= 3 + x \\
16(x^2 + y^2) &= 9 + 6x + x^2.
\end{aligned}
$$

Thus, we obtain $15x^2 + 16y^2 - 6x - 9 = 0$ is an ellipse.

27. Multiply by $3 + 9\cos\theta$. Then

$$
\begin{aligned}
3r + 9r\cos\theta &= 1 \\
3r &= 1 - 9x \\
9(x^2 + y^2) &= 1 - 18x + 81x^2.
\end{aligned}
$$

Thus, $72x^2 - 9y^2 - 18x + 1 = 0$ is a hyperbola.

29. Multiply by $6 - \sin\theta$. Then

$$
\begin{aligned}
6r - r\sin\theta &= 2 \\
6r &= 2 + y \\
36(x^2 + y^2) &= 4 + 4y + y^2.
\end{aligned}
$$

Thus, we get $36x^2 + 35y^2 - 4y - 4 = 0$ is an ellipse.

31. Since $r = \dfrac{\frac{1}{3} \cdot 12}{1 + \frac{1}{3}\cos\theta}$, $e = \dfrac{1}{3}$. By letting $\theta = 0, \pi$, we get the vertices in polar coordinates, $(3, 0)$ and $(6, \pi)$. The length of the major axis is $2a = 9$. So $a = \dfrac{9}{2}$. Recall, if $2c$ is the distance between the foci then $e = \dfrac{c}{a}$. So, $\dfrac{1}{3} = \dfrac{c}{9/2}$ and $2c = 3$. Thus, the foci are $(0,0)$ and $(3, \pi)$.

33. By letting $\theta = 0, \pi$, we obtain the vertices $(2, 0)$ and $(2/3, \pi)$ in polar coordinates of the ellipse. So, one-half the length of the major axis is $a = \dfrac{4}{3}$. Since the distance c between the pole and the vertex $(2/3, \pi)$ is $c = \dfrac{2}{3}$, the eccentricity is

$$
e = \frac{c}{a} = \frac{2/3}{4/3} = \frac{1}{2}.
$$

For Thought

1. False, t is the parameter.

2. True, graphs of parametric equations are sketched in a rectangular coordinate system in this book.

3. True, since $2x = t$ and $y = 2t + 1 = 2(2x) + 1 = 4x + 1$.

4. False, it is a circle of radius 1.

5. True, since if $t = \dfrac{1}{3}$ then $x = 3\left(\dfrac{1}{3}\right) + 1 = 2$ and $y = 6\left(\dfrac{1}{3}\right) - 1 = 1$.

6. False, for if $w^2 - 3 = 1$ then $w = \pm 2$ and this does not satisfy $-2 < w < 2$.

7. True, since x and y take only positive values.

8. True

9. False, since e^t is non-negative while $\ln(t)$ can be negative.

10. True

5.6 Exercises

1. If $t = 0$, then $x = 4(0) + 1 = 1$ and $y = 0 - 2 = -2$. If $t = 1$, then $x = 4(1) + 1 = 5$ and $y = 1 - 2 = -1$.

If $x = 7$, then $7 = 4t + 1$. Solving for t, we get $t = 1.5$. Substitute $t = 1.5$ into $y = t - 2$. Then $y = 1.5 - 2 = -0.5$.

If $y = 1$, then $1 = t - 2$. Solving for t, we get $t = 3$. Consequently $y = 4(3) + 1 = 13$.

We tabulate the results as follows.

t	x	y
0	1	-2
1	5	-1
1.5	7	-0.5
3	13	1

3. If $t = 1$, then $x = 1^2 = 1$ and $y = 3(1) - 1 = 2$. If $t = 2.5$, then $x = (2.5)^2 = 6.25$ and $y = 3(2.5) - 1 = 6.5$.

If $x = 5$, then $5 = t^2$ and $t = \sqrt{5}$. Consequently, $y = 3\sqrt{5} - 1$.

If $y = 11$, then $11 = 3t - 1$. Solving for t, we get $t = 4$. Consequently $x = 4^2 = 16$.

If $x = 25$, then $25 = t^2$ and $t = 5$. Consequently, $y = 3(5) - 1 = 14$.

We tabulate the results as follows.

t	x	y
1	1	2
2.5	6.25	6.5
$\sqrt{5}$	5	$3\sqrt{5}-1$
4	16	11
5	25	14

5.

Some points are given by
t	x	y
0	-2	3
4	10	7

The domain is $[-2, 10]$ and the range is $[3, 7]$.

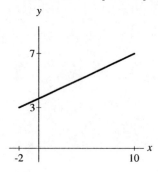

7.

Some points are given by
t	x	y
0	-1	0
2	1	4

The domain is $(-\infty, \infty)$ and the range is $[0, \infty)$.

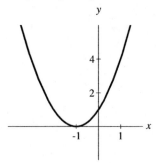

9. A few points are approximated by

w	x	y
0.2	0.4	0.9
0.8	0.9	0.4

The domain is $(0, 1)$ and the range is $(0, 1)$.

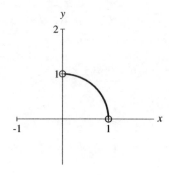

11. A circle of radius 1 and centered at the origin. The domain is $[-1, 1]$ and the the range is $[-1, 1]$.

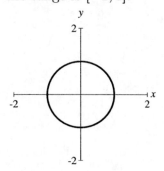

13. Since $t = \dfrac{x+5}{4}$, we obtain

$$y = 3 - 4\left(\frac{x+5}{4}\right) = -x - 2 \text{ or } x + y = -2.$$

The graph is a straight line with domain $(-\infty, \infty)$ and range $(-\infty, \infty)$.

15. Since $x^2 + y^2 = 16\sin^2(3t) + 16\cos^2(3t) = 16$, the graph is a circle with radius 4 and with center at the origin.

Domain $[-4, 4]$ and range $[-4, 4]$

17. Since $t = 4x$, we find $y = e^{4x}$ and the graph is an exponential graph.

Domain $(-\infty, \infty)$ and range $(0, \infty)$

19. $y = 2x + 3$ represents the graph of a straight line

Domain $(-\infty, \infty)$ and range $(-\infty, \infty)$

21. An equation (in terms of x and y) of the line through $(2, 3)$ and $(5, 9)$ is $y = 2x - 1$.

An equation (in terms of t and x) of the line through $(0, 2)$ and $(2, 5)$ is $x = \dfrac{3}{2}t + 2$.

Parametric equations are $x = \dfrac{3}{2}t + 2$ and

$y = 2\left(\dfrac{3}{2}t + 2\right) - 1 = 3t + 3$ where $0 \le t \le 2$.

23. $x = 2\cos t,\ y = 2\sin t,\ \pi < t < \dfrac{3\pi}{2}$

25. $x = 3,\ y = t,\ -\infty < t < \infty$

27. Since $x = r\cos\theta$ and $y = r\sin\theta$, we get
$x = 2\sin t\cos t$ and $y = 2\sin t\sin t$
where $0 \le t \le 2\pi$.

Equivalently, parametric equations are
$x = \sin 2t$ and $y = 2\sin^2 t$ where
$0 \le t \le 2\pi$.

29. For $-\pi \le t \le \pi$, one obtains the given graph
(for a larger range of values for t, more points
are filled and the graph would be different)

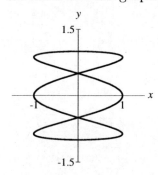

31. For $-15 \le t \le 15$, one finds

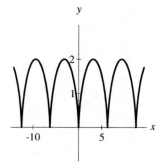

33. For $-10 \le t \le 10$, one obtains

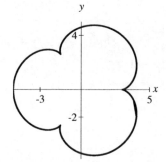

35. A graph of the parametric equations

$$x = 150\sqrt{3}\,t$$

and

$$y = -16t^2 + 150t + 5$$

for $0 \le t \le 10$ is given

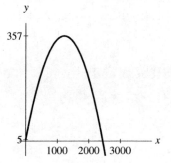

37. Solving $y = -16t^2 + 150t + 5 = 0$, one finds

$$\frac{-150 \pm \sqrt{150^2 - 4(-16)(5)}}{-32} \approx 9.41,\ -0.03$$

The arrow is in the air for 9.4 seconds.

Review Exercises

1. If $y = 0$, then by factoring we get
$0 = (x + 6)(x - 2)$ and the x-intercepts are
$(2, 0)$ and $(-6, 0)$.

If $x = 0$, then $y = -12$ and the y-intercept is
$(0, -12)$.

By completing the square, we obtain
$y = (x + 2)^2 - 16$, vertex $(h, k) = (-2, -16)$,
and the axis of symmetry is $x = -2$.

Since $p = \dfrac{1}{4a} = \dfrac{1}{4}$, the focus is

$(h, k + p) = (-2, -63/4)$ and
directrix is $y = k - p$ or $y = -65/4$.

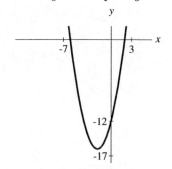

3. If $y = 0$, then by factoring we find $0 = x(6-2x)$ and the x-intercepts are $(0,0)$ and $(3,0)$.

If $x = 0$, then $y = 0$ and the y-intercept is $(0,0)$.

By completing the square, one gets $y = -2(x - 3/2)^2 + 9/2$, with vertex $(h, k) = (3/2, 9/2)$, and axis of symmetry is $x = 3/2$.

Since $p = \dfrac{1}{4a} = -\dfrac{1}{8}$, the focus is $(h, k + p) = (3/2, 35/8)$ and the directrix is $y = k - p$ or $y = 37/8$.

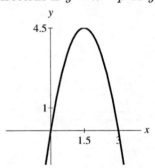

5. By completing the square, we have $x = (y+2)^2 - 10$. If $x = 0$, then $y+2 = \pm\sqrt{10}$ and the y-intercepts are $(0, -2 \pm \sqrt{10})$.

If $y = 0$, then $x = -6$ and the x-intercept is $(-6, 0)$. Since $x = (y + 2)^2 - 10$ is of the form $x = a(y - h)^2 + k$, the vertex is $(k, h) = (-10, -2)$, and the axis of symmetry is $y = -2$.

Since $p = \dfrac{1}{4a} = \dfrac{1}{4}$, the focus is

$$(k + p, h) = (-39/4, -2)$$

and directrix is $x = k - p$ or $x = -41/4$.

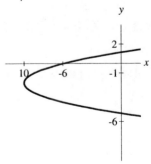

7. Since $c = \sqrt{a^2 - b^2} = \sqrt{36 - 16} = 2\sqrt{5}$, the foci are $(0, \pm 2\sqrt{5})$

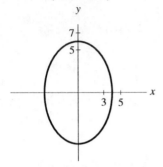

9. Since $c = \sqrt{a^2 - b^2} = \sqrt{24 - 8} = 4$, the foci are $(1, 1 \pm 4)$, or $(1, 5)$ and $(1, -3)$.

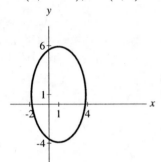

11. Since $c = \sqrt{a^2 - b^2} = \sqrt{10 - 8} = \sqrt{2}$, the foci are $(1, -3 \pm \sqrt{2})$.

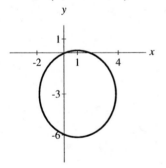

13. center $(0, 0)$, radius 9

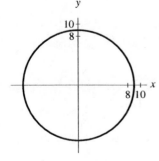

15. center $(-1, 0)$, radius 2

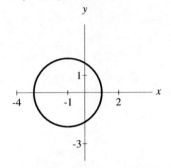

17. Completing the square, we have

$$x^2 + 5x + \frac{25}{4} + y^2 = -\frac{1}{4} + \frac{25}{4}$$
$$\left(x + \frac{5}{2}\right)^2 + y^2 = 6,$$

and so center is $(-5/2, 0)$ and radius is $\sqrt{6}$.

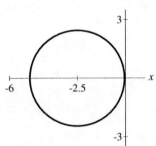

19. $x^2 + (y + 4)^2 = 9$

21. $(x + 2)^2 + (y + 7)^2 = 6$

23. Since $c = \sqrt{a^2 + b^2} = \sqrt{8^2 + 6^2} = 10$,

foci $(\pm 10, 0)$, asymptotes $y = \pm\frac{b}{a} = \pm\frac{3}{4}x$

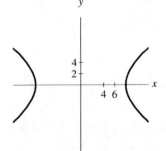

25. Since $c = \sqrt{a^2 + b^2} = \sqrt{8^2 + 4^2} = 4\sqrt{5}$,

the foci are $(4, 2 \pm 4\sqrt{5})$. Solving

for y in $y - 2 = \pm\frac{8}{4}(x - 4)$,

asymptotes are $y = 2x - 6$ and $y = -2x + 10$.

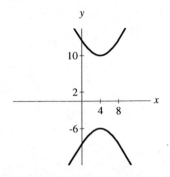

27. Completing the square, we have

$$\begin{aligned}
x^2 - 4x - 4(y^2 - 8y) &= 64 \\
x^2 - 4x + 4 - 4(y^2 - 8y + 16) &= 64 + 4 - 64 \\
(x - 2)^2 - 4(y - 4)^2 &= 4 \\
\frac{(x - 2)^2}{4} - (y - 4)^2 &= 1,
\end{aligned}$$

and so $c = \sqrt{a^2 + b^2} = \sqrt{2^2 + 1^2} = \sqrt{5}$, and

the foci are $(2 \pm \sqrt{5}, 4)$. Solving for y

in $y - 4 = \pm\frac{1}{2}(x - 2)$, we get that the

asymptotes are $y = \frac{1}{2}x + 3$ and $y = -\frac{1}{2}x + 5$.

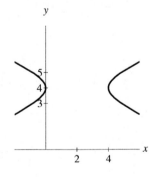

29. Hyperbola

31. Ellipse

33. Parabola

35. Hyperbola

37. $x^2 + y^2 = 4$ is a circle

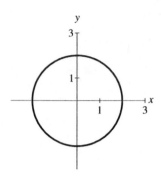

39. Since $4y = x^2 - 4$, $y = \dfrac{1}{4}x^2 - 1$.

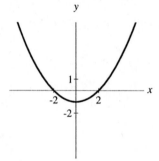

41. Since $x^2 + 4y^2 = 4$, $\dfrac{x^2}{4} + y^2 = 1$.

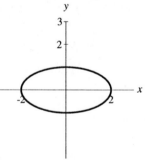

43. Since $x^2 - 4x + 4 + 4y^2 = 4$, we find

$$(x-2)^2 + 4y^2 = 4 \text{ and } \dfrac{(x-2)^2}{4} + y^2 = 1.$$

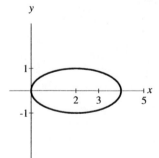

45. Since the vertex is midway between the focus $(1,3)$ and directrix $x = \dfrac{1}{2}$, the vertex is $\left(\dfrac{3}{4}, 3\right)$ and $p = \dfrac{1}{4}$. Since

$a = \dfrac{1}{4p} = 1$, parabola is given by

$$x = (y-3)^2 + \dfrac{3}{4}.$$

47. From the foci and vertices one gets $c = 4$ and $a = 6$, respectively. Since $b^2 = a^2 - c^2 = 36 - 16 = 20$, the

ellipse is given by $\dfrac{x^2}{36} + \dfrac{y^2}{20} = 1$.

49. Radius is $\sqrt{(-1-1)^2 + (-1-3)^2} = \sqrt{20}$. Equation is $(x-1)^2 + (y-3)^2 = 20$.

51. From the foci and x-intercepts one gets $c = 3$ and $a = 2$, respectively. Since $b^2 = c^2 - a^2 = 9 - 4 = 5$, the

hyperbola is given by $\dfrac{x^2}{4} - \dfrac{y^2}{5} = 1$.

53. Since the center of the circle is $(-2,3)$ and the raidus is 3, we have $(x+2)^2 + (y-3)^2 = 9$.

55. Note, the center of the ellipse is $(-2,1)$. Using the lengths of the major axis, we get $a = 3$ and $b = 1$. Thus, an equation is

$$\dfrac{(x+2)^2}{9} + (y-1)^2 = 1.$$

57. Note, the center of the hyperbola is $(2,1)$. By using the fundamental rectangle, we find $a = 3$ and $b = 2$. Thus, an equation is

$$\dfrac{(y-1)^2}{9} - \dfrac{(x-2)^2}{4} = 1.$$

59. $(5\cos 60°, 5\sin 60°) = \left(\dfrac{5}{2}, \dfrac{5\sqrt{3}}{2}\right)$

61. $(\sqrt{3}\cos 100°, \sqrt{3}\sin 100°) \approx (-0.3, 1.7)$

63. Note $r = \sqrt{(-2)^2 + (-2\sqrt{3})^2} = \sqrt{16} = 4$. Since $\tan\theta = \sqrt{3}$ and the terminal side of θ goes through $(-2, -2\sqrt{3})$, we have

$$\theta = 4\pi/3. \text{ Then } (r, \theta) = \left(4, \dfrac{4\pi}{3}\right).$$

65. Note $r = \sqrt{2^2 + (-3)^2} = \sqrt{13}$.
Since $\theta = \tan^{-1}(-3/2) \approx -0.98$,
we have $(r, \theta) \approx \left(\sqrt{13}, -0.98\right)$.

67. Circle centered at $(r, \theta) = (1, -\pi/2)$

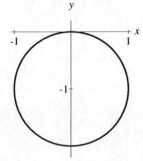

69. four-leaf rose

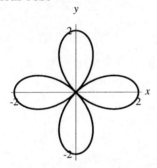

71. Limacon $r = 500 + \cos\theta$

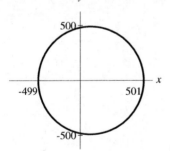

73. Horizontal line $y = 1$

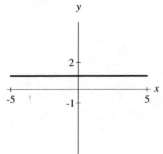

75. Since $r = \dfrac{1}{\sin\theta + \cos\theta}$, we obtain

$$
\begin{aligned}
r\sin\theta + r\cos\theta &= 1 \\
y + x &= 1.
\end{aligned}
$$

77. $x^2 + y^2 = 25$

79. Since $y = 3$, we find $r\sin\theta = 3$ and $r = \dfrac{3}{\sin\theta}$.

81. $r = 7$

83. Parabola with directrix $y = 3$

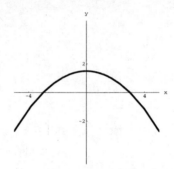

85. When we rewrite the equation as

$$
r = \dfrac{\frac{1}{2} \cdot 4}{1 + \frac{1}{2}\cos\theta}
$$

we obtain an ellipse with directrix $x = 4$.

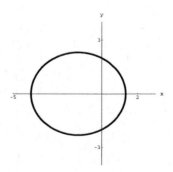

87. Since $r = \dfrac{2 \cdot \frac{1}{6}}{1 - 2\cos\theta}$, we have a hyperbola

with directrix $x = -\dfrac{1}{6}$

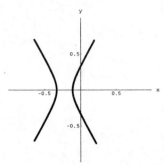

89. $r = \dfrac{3}{1 + \sin\theta}$

91. Since $r = \dfrac{3(6)}{1 - 3\cos\theta}$, we find $r = \dfrac{18}{1 - 3\cos\theta}$.

93. Note, $r = \dfrac{\frac{1}{3} \cdot 9}{1 + \frac{1}{3}\sin\theta}$. Then $r = \dfrac{9}{3 + \sin\theta}$.

95. The boundary points are given by

t	x	y
0	0	3
1	3	2

Note, the boundary points do not lie on the graph as shown in the next column.

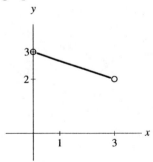

97. The graph is a quarter of a circle.

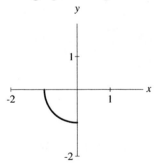

99. The equation is of the form $\dfrac{x^2}{100^2} - \dfrac{y^2}{b^2} = 1$.

Since the graph passes through $(120, 24\sqrt{11})$, we get

$$\frac{120^2}{100^2} - \frac{(24\sqrt{11})^2}{b^2} = 1$$

$$1.44 - \frac{6336}{b^2} = 1$$

$$b^2 = \frac{6336}{0.44}$$

$$b^2 = 120^2.$$

The equation is

$$\frac{x^2}{100^2} - \frac{y^2}{120^2} = 1.$$

101. Note $c = 30$ and $a = 34$. Then an equation we can use is of the form $\dfrac{x^2}{34^2} + \dfrac{y^2}{b^2} = 1$.

Since $b^2 = a^2 - c^2 = 34^2 - 30^2 = 16^2$, the equation is

$$\frac{x^2}{34^2} + \frac{y^2}{16^2} = 1.$$

To find h, let $x = 32$.

$$\frac{32^2}{34^2} + \frac{y^2}{16^2} = 1$$

$$y^2 = \left(1 - \frac{32^2}{34^2}\right)16^2$$

$$y \approx 5.407.$$

Thus, $h = 2y \approx 10.81$ feet.

A.1 Exercises

1. 64

3. -49

5. $(-4)(-4) = 16$

7. $\dfrac{1}{3^4} = \dfrac{1}{81}$

9. $\dfrac{1}{6} + \dfrac{1}{5} = \dfrac{5}{30} + \dfrac{6}{30} = \dfrac{11}{30}$

11. $\dfrac{6^3}{3^2} = 24$

13. $2^3 = 8$

15. $-6x^{11}y^{11}$

17. $y^5 + 2y^5 = 3y^5$

19. $-4x^6$

21. $\dfrac{(-2)^3(x^2)^3}{27} = \dfrac{-8x^6}{27}$

23. $3x^4$

25. $\dfrac{25}{y^4}$

27. $\dfrac{1}{6}x^0 y^{-3} = \dfrac{1}{6y^3}$

29. $\left(\dfrac{n^2}{2}\right)^2 = \dfrac{n^4}{4}$

31. -3

33. 8

35. -4

37. 81

39. $(8^{1/3})^{-4} = \dfrac{1}{16}$

41. $x^2 y^{1/2}$

43. $6a^{1/2+1} = 6a^{3/2}$

45. $3a^{1/2-1/3} = 3a^{1/6}$

47. $a^{2+1/3}b^{1/2+1/2} = a^{7/3}b$

49. $\dfrac{x^2 y}{z^3}$

51. 30

53. -2

55. $-\dfrac{\sqrt[3]{8}}{\sqrt[3]{1000}} = -\dfrac{1}{5}$

57. $\sqrt[4]{(2^4)^3} = \sqrt[4]{2^{12}} = 2^{12\cdot(1/4)} = 2^3 = 8$

59. $10^{2/3} = \sqrt[3]{10^2}$

61. $\dfrac{3}{y^{3/5}} = \dfrac{3}{\sqrt[5]{y^3}}$

63. $x^{-1/2}$

65. $(x^3)^{1/5} = x^{3/5}$

67. $4x$

69. $2y^3$

71. $\dfrac{\sqrt{xy}}{10}$

73. $\dfrac{-2a}{b^{15\cdot(1/3)}} = -\dfrac{2a}{b^5}$

75. $\sqrt{4(7)} = 2\sqrt{7}$

77. $\dfrac{1}{\sqrt{5}} \cdot \dfrac{\sqrt{5}}{\sqrt{5}} = \dfrac{\sqrt{5}}{5}$

79. $\dfrac{\sqrt{x}}{\sqrt{8}} \cdot \dfrac{\sqrt{2}}{\sqrt{2}} = \dfrac{\sqrt{2x}}{\sqrt{16}} = \dfrac{\sqrt{2x}}{4}$

81. $\sqrt[3]{8\cdot 5} = 2\sqrt[3]{5}$

83. $\sqrt[3]{-250x^4} = \sqrt[3]{(-125)(2)x^3 x} = -5x\sqrt[3]{2x}$

85. $\sqrt[3]{\dfrac{1}{2}} \cdot \sqrt[3]{\dfrac{4}{4}} = \sqrt[3]{\dfrac{4}{8}} = \dfrac{\sqrt[3]{4}}{2}$

87. $9 - 2\sqrt{6}$

89. $2\sqrt{2} + 2\sqrt{5} - 2\sqrt{3}$

91. $-10\sqrt{18} = -30\sqrt{2}$

93. $12(5a) = 60a$

95. $\sqrt{\dfrac{9}{a^3}} = \dfrac{3}{a\sqrt{a}} \cdot \dfrac{\sqrt{a}}{\sqrt{a}} = \dfrac{3\sqrt{a}}{a^2}$

97. $2x\sqrt{5x} + 3x\sqrt{5x} = 5x\sqrt{5x}$

A.2 Exercises

1. Degree 3 trinomial with leading coefficient 1

3. Degree 2 binomial with leading coefficient -3

5. Degree 0 monomial with leading coefficient 79

7. $P(-2) = 4 + 6 + 2 = 12$

9. $M(-3) = 27 + 45 + 3 + 2 = 77$

11. $8x^2 + 3x - 1$

13. $4x^2 - 3x - 9x^2 + 4x - 3 = -5x^2 + x - 3$

15. $4ax^3 - a^2x - 5a^2x^3 + 3a^2x - 3 = $
$(-5a^2 + 4a)x^3 + 2a^2x - 3$

17. $2x - 1$　　**19.** $3x^2 - 3x - 6$

21. $-18a^5 + 15a^4 - 6a^3$

23. $(3b^2 - 5b + 2)b - (3b^2 - 5b + 2)3 = $
$3b^3 - 5b^2 + 2b - 9b^2 + 15b - 6 = $
$3b^3 - 14b^2 + 17b - 6$

25. $2x(4x^2 + 2x + 1) - (4x^2 + 2x + 1) = $
$8x^3 + 4x^2 + 2x - 4x^2 - 2x - 1 = $
$8x^3 - 1$

27. $x(x^2 - 5x + 25) + 5(x^2 - 5x + 25) = $
$x^3 - 5x^2 + 25x + 5x^2 - 25x + 125 = $
$x^3 + 125$

29. $(x - 4)z + (x - 4)3 = xz - 4z + 3x - 12$

31. $a(a^2 + ab + b^2) - b(a^2 + ab + b^2) = $
$a^3 + a^2b + ab^2 - a^2b - ab^2 - b^3 = a^3 - b^3$

33. $a^2 - 2a + 9a - 18 = a^2 + 7a - 18$

35. $2y^2 + 18y - 3y - 27 = 2y^2 + 15y - 27$

37. $4x^2 + 18x - 18x - 81 = 4x^2 - 81$

39. $6x^4 + 10x^2 + 12x^2 + 20 = 6x^4 + 22x^2 + 20$

41. $4x^2 + 10x + 10x + 25 = 4x^2 + 20x + 25$

43. $(3x)^2 + 2(3x)(5) + (5)^2 = 9x^2 + 30x + 25$

45. $(x^2)^2 - 3^2 = x^4 - 9$

47. $(\sqrt{2})^2 - 5^2 = -23$

49. $(3x^3)^2 - 2(3x^3)(4) + 4^2 = 9x^6 - 24x^3 + 16$

51. $4x^2y^2 - 20xy + 25$

53. $\dfrac{\sqrt{10}}{\sqrt{5} - 2} \cdot \dfrac{\sqrt{5} + 2}{\sqrt{5} + 2} = \dfrac{\sqrt{50} + 2\sqrt{10}}{5 - 4} = 5\sqrt{2} + 2\sqrt{10}$

55. $\dfrac{\sqrt{6}}{6 + \sqrt{3}} \cdot \dfrac{6 - \sqrt{3}}{6 - \sqrt{3}} = \dfrac{6\sqrt{6} - 3\sqrt{2}}{33} = \dfrac{2\sqrt{6} - \sqrt{2}}{11}$

57. Quotient $x + 3$, remainder 0 since

$$
\begin{array}{r}
x + 3 \\
x + 3 \enclose{longdiv}{x^2 + 6x + 9} \\
\underline{x^2 + 3x} \\
3x + 9 \\
\underline{3x + 9} \\
0
\end{array}
$$

59. Quotient $a^2 + a + 1$, remainder 0 since

$$
\begin{array}{r}
a^2 + a + 1 \\
a - 1 \enclose{longdiv}{a^3 - 1} \\
\underline{a^3 - a^2} \\
a^2 - 1 \\
\underline{a^2 - a} \\
a - 1 \\
\underline{a - 1} \\
0
\end{array}
$$

61. Quotient $x + 5$, remainder 13 since

$$
\begin{array}{r}
x + 5 \\
x - 2 \enclose{longdiv}{x^2 + 3x + 3} \\
\underline{x^2 - 2x} \\
5x + 3 \\
\underline{5x - 10} \\
13
\end{array}
$$

63. Quotient $2x - 6$, remainder 13 since

$$
\begin{array}{r}
2x - 6 \\
x + 3 \enclose{longdiv}{2x^2 + 0x - 5} \\
\underline{2x^2 + 6x} \\
-6x - 5 \\
\underline{-6x - 18} \\
13
\end{array}
$$

65. Quotient $x^2 + x + 1$, remainder 0 since

$$
\begin{array}{r}
x^2 + x + 1 \\
x - 3 \enclose{longdiv}{x^3 - 2x^2 - 2x - 3} \\
\underline{x^3 - 3x^2} \\
x^2 - 2x \\
\underline{x^2 - 3x} \\
x - 3 \\
\underline{x - 3} \\
0
\end{array}
$$

67. Quotient $3x - 5$, remainder 7 since

$$
\begin{array}{r}
3x - 5 \\
2x + 1 \enclose{longdiv}{6x^2 - 7x + 2} \\
\underline{6x^2 + 3x} \\
-10x + 2 \\
\underline{-10x - 5} \\
7
\end{array}
$$

69. Quotient $2x + 3$, remainder x since

$$
\begin{array}{r}
2x + 3 \\
x^2 - 4 \enclose{longdiv}{2x^3 + 3x^2 - 7x - 12} \\
\underline{2x^3 \qquad - 8x} \\
3x^2 + x - 12 \\
\underline{3x^2 \qquad - 12} \\
x
\end{array}
$$

71. Quotient $x + 3$, remainder $-2x$ since

$$
\begin{array}{r}
x + 3 \\
x^2 - x - 2 \enclose{longdiv}{x^3 + 2x^2 - 7x - 6} \\
\underline{x^3 - x^2 - 2x} \\
3x^2 - 5x - 6 \\
\underline{3x^2 - 3x - 6} \\
-2x
\end{array}
$$

73. $x^2 + 2x - 24$

75. $2a^{10} - 3a^5 - 27$

77. $-y - 9$

79. $w^2 + 8w + 16$

81. $3y^5 - 9xy^2$

83. $2b - 1$

85. $9w^4 - 12w^2n + 4n^2$

A.3 Exercises

1. $6x^2(x - 2),\ -6x^2(-x + 2)$

3. $ax(-x^2 + 5x - 6), -ax(x^2 - 5x + 6)$

5. $1(m - n), -1(n - m)$

7. $x^2(x + 2) + 5(x + 2) = (x^2 + 5)(x + 2)$

9. $y^2(y - 1) - 3(y - 1) = (y^2 - 3)(y - 1)$

11. $ady + d - awy - w = d(ay + 1) - w(ay + 1) =$
$(ay + 1)(d - w)$

13. $x^2y^2 - ay^2 - (bx^2 - ab) = y^2(x^2 - a) - b(x^2 - a)$
$= (x^2 - a)(y^2 - b)$

15. $(x + 2)(x + 8)$

17. $(x - 6)(x + 2)$

19. $(m - 2)(m - 10)$

21. $(t - 7)(t + 12)$

23. $(2x + 1)(x - 4)$

25. $(4x + 1)(2x - 3)$

27. $(3y + 5)(2y - 1)$

29. $(3b + 2)(4b + 3)$

31. $(t - u)(t + u)$

33. $(t + 1)^2$

35. $(2w - 1)^2$

37. $(3zx + 4)^2$

39. $(t - u)(t^2 + ut + u^2)$

41. $(a - 2)(a^2 + 2a + 4)$

43. $(3y + 2)(9y^2 - 6y + 4)$

45. $(3xy^2)^3 - (2z^3)^3 =$
$(3xy^2 - 2z^3)(9x^2y^4 + 6xy^2z^3 + 4z^6)$

47. $-3x^3 + 27x = -3x(x^2 - 9) = -3x(x - 3)(x + 3)$

49. $2t(8t^3 + 27w^3) = 2t(2t + 3w)(4t^2 - 6tw + 9w^2)$

51. $a^3 + a^2 - 4a - 4 = a^2(a + 1) - 4(a + 1) =$
$(a + 1)(a^2 - 4) = (a + 1)(a + 2)(a - 2)$

53. $x^4 - 2x^3 - 8x + 16 = x^3(x - 2) - 8(x - 2) =$
$(x - 2)(x^3 - 8) = (x - 2)^2(x^2 + 2x + 4)$

55. $-2x(18x^2 - 9x - 2) = -2x(6x + 1)(3x - 2)$

57. $a^7 - a^6 - 64a + 64 = a^6(a - 1) - 64(a - 1) =$
$(a^6 - 64)(a - 1) = (a^3 - 8)(a^3 + 8)(a - 1) =$
$(a - 2)(a^2 + 2a + 4)(a + 2)(a^2 - 2a + 4)(a - 1)$

59. $-(3x + 5)(2x - 3)$

A.4 Exercises

1. $\{x \mid x \neq -2\}$

3. $\{x \mid x \neq 4, -2\}$

5. $\{x \mid x \neq \pm 3\}$

7. All real numbers since $x^2 + 3 \neq 0$ for any real number x.

9. $\dfrac{3(x-3)}{(x-3)(x+2)} = \dfrac{3}{x+2}$

11. $\dfrac{10a - 8b}{12b - 15a} = \dfrac{2(5a - 4b)}{-3(5a - 4b)} = -\dfrac{2}{3}$

13. $\dfrac{a^3 b^6}{a^2 b^3 - a^4 b^2} = \dfrac{a^3 b^6}{a^2 b^2 (b - a^2)} = \dfrac{ab^4}{b - a^2}$

15. $\dfrac{y^2 z}{x^3}$

17. $\dfrac{a^3 - b^3}{a^2 - b^2} = \dfrac{(a-b)(a^2 + ab + b^2)}{(a-b)(a+b)} = \dfrac{a^2 + ab + b^2}{a + b}$

19. $\dfrac{a(b+3) - y(b+3)}{(a-y)^2} = \dfrac{(a-y)(b+3))}{(a-y)^2} = \dfrac{b+3}{a - y} =$

21. $\dfrac{2a}{3b^2} \cdot \dfrac{9b}{14a^2} = \dfrac{1}{b} \cdot \dfrac{3}{7a} = \dfrac{3}{7ab}$

23. $\dfrac{12a}{7} \cdot \dfrac{49}{2a^3} = \dfrac{42}{a^2}$

25. $\dfrac{(a-3)(a+3)}{3(a-2)} \cdot \dfrac{(a-2)(a+2)}{(a-3)(a+2)} = \dfrac{a+3}{3}$

27. $\dfrac{(x-y)(x+y)}{9} \cdot \dfrac{9(2)}{(x+y)^2} = \dfrac{2x - 2y}{x + y}$

29. $\dfrac{(x-y)(x+y)}{-3xy} \cdot \dfrac{3xy(2xy^2)}{-2(x-y)} = x^2 y^2 + xy^3$

31. $\dfrac{x(w-1)}{x^2} \cdot \dfrac{2}{(1-w)(1+w)} = \dfrac{-2}{x(1+w)}$

$= \dfrac{-2}{x + wx}$

33. $\dfrac{16a}{12a^2}$

35. $\dfrac{x-5}{x+3} \cdot \dfrac{x-3}{x-3} = \dfrac{x^2 - 8x + 15}{x^2 - 9}$

37. $\dfrac{x}{x+5} \cdot \dfrac{x+1}{x+1} = \dfrac{x^2 + x}{x^2 + 6x + 5}$

39. $\dfrac{t}{2t+2} \cdot \dfrac{t+1}{t+1} = \dfrac{t^2 + t}{2t^2 + 4t + 2}$

41. $12a^2 b^3$

43. Since $3a + 3b = 3(a+b)$ and $2a + 2b = 2(a+b)$, the LCD is $6(a+b)$.

45. Since $x^2 + 5x + 6 = (x+3)(x+2)$ and $x^2 - x - 6 = (x-3)(x+2)$, the LCD is $(x+3)(x-3)(x+2)$.

47. $\dfrac{3(3)}{2x(3)} + \dfrac{x}{6x} = \dfrac{9 + x}{6x}$

49.

$\dfrac{(x+3)(x+1)}{(x-1)(x+1)} - \dfrac{(x+4)(x-1)}{(x+1)(x-1)} =$

$\dfrac{x^2 + 4x + 3}{(x-1)(x+1)} - \dfrac{x^2 + 3x - 4}{(x+1)(x-1)} =$

$\dfrac{x+7}{(x+1)(x-1)}$

51. $\dfrac{3a}{a} + \dfrac{1}{a} = \dfrac{3a + 1}{a}$

53. $\dfrac{(t-1)(t+1)}{t+1} - \dfrac{1}{t+1} = \dfrac{t^2 - 2}{t + 1}$

55.

$\dfrac{x}{(x+2)(x+1)} + \dfrac{x-1}{(x+3)(x+2)} =$

$\dfrac{x(x+3)}{(x+2)(x+1)(x+3)} +$

$\dfrac{(x-1)(x+1)}{(x+3)(x+2)(x+1)} =$

$\dfrac{2x^2 + 3x - 1}{(x+1)(x+2)(x+3)}$

57.

$\dfrac{1}{x-3} - \dfrac{5}{-2(x-3)} = \dfrac{2}{2(x-3)} - \dfrac{-5}{2(x-3)}$

$= \dfrac{7}{2x - 6}$

59.

$\dfrac{y^2}{(x-y)(x^2 + xy + y^2)} +$

$\dfrac{(x+y)(x-y)}{(x^2 + xy + y^2)(x-y)} =$

$\dfrac{y^2}{(x-y)(x^2 + xy + y^2)} +$

$$\frac{x^2 - y^2}{(x^2 + xy + y^2)(x - y)} =$$

$$\frac{x^2}{x^3 - y^3}$$

61.

$$\frac{x - 2}{(2x - 3)(x + 5)} - \frac{x + 1}{(2x - 3)(x - 1)} =$$

$$\frac{(x - 2)(x - 1)}{(2x - 3)(x + 5)(x - 1)} -$$

$$\frac{(x + 1)(x + 5)}{(2x - 3)(x + 5)(x - 1)} =$$

$$\frac{x^2 - 3x + 2}{(2x - 3)(x + 5)(x - 1)} -$$

$$\frac{x^2 + 6x + 5}{(2x - 3)(x + 5)(x - 1)} =$$

$$\frac{-9x - 3}{(2x - 3)(x + 5)(x - 1)} =$$